Killing Waters

The Great West Virginia Flood Of 1985

Compiled and edited by
Bob Teets and Shelby Young

About this book:

There can never be enough written, enough photos taken, enough said about what happened to the people of West Virginia on the night of November 4, 1985. Unprecedented floodwaters swept over a state already burdened with grief from the disasters at Farmington Number Nine Mine, Willow Island, Buffalo Creek and scores of others. The Great West Virginia Flood of 1985 added to the grief.

But as with other calamities, West Virginians faced this catastrophe with courage, dignity and resolve.

And it is to these Mountaineers that we dedicate this book.

Bob Teets,
Editor/Publisher
Preston County News
Shelby Young,
Managing Editor,
Clay County Free Press

805 East State Ave.
Terra Alta, W. Va. 26764
ISBN 0-87012-48-1

McClain Printing Company
Parsons, West Virginia

Contents include excerpts from these excellent newspapers:

The Morgan Messenger, Berkeley Springs
The Mineral Daily News-Tribune, Keyser
The Piedmont Herald, Piedmont
Dominion Post, Morgantown
Grant County Press, Petersburg
Hampshire Review, Romney
The Inter-Mountain, Elkins
Preston County News, Terra Alta
The Tyler Star News and Tyler County Journal, Sistersville
Herald Dispatch, Huntington
Spirit of Jefferson Advocate, Charles Town
Wirt County Journal, Elizabeth
The Glenville Democrat and Pathfinder, Glenville
The Preston County Journal, Kingwood
The Pocahontas Times, Marlinton
The Shinnston News, Shinnston
The Gulf Times, Sophia
The Herald Record, West Union
Clarksburg Exponent-Telegram, Clarksburg
Times West Virginian, Fairmont
West Virginia Daily News, Lewisburg
The Intelligencer, Wheeling
Record Delta, Buckhannon
Clay County Free Press, Clay
The Pendleton Times, Franklin
The Calhoun Cronicle and Grantsville News, Grantsville
The Star-Independent, Lewisburg
The Moorefield Examiner, Moorefield
Parsons Advocate, Parsons
The News Leader, Richwood
St. Marys Oracle and Pleasants County Leader
Braxton Citizens' News, Sutton
The Webster Echo and Republican, Webster Springs
Weston Democrat, Weston
Mt. Statesman, Grafton
The Evening Journal, Martinsburg
Parkersburg News and Sentinel, Parkersburg

Thanks to these friends:

Bill Childress, WVPA
George and Mariwyn Smith, McClain Printing
The Staff of the Preston County News
Laura Young
Cathy Teets
Patti Grose
Greg Hunter
D. Clayton Booth
Barbara Davis
Jim Teets
Adam Kelly
Eldora Nuzum
Dave Owen
Cecil B. Highland, Jr.
Randall Strother
Robert Hammond
Tina Bolyard
Don Hatfield
Walter D. Tetrick
Frank Spicer
Page Burdette
James D. Hoffman
Francis Hollendonner
Bob DeFrancis
G. Ogden Nutting
Phil Forney
J. Warren Buzzard
Don Ranch
Clinton N. Nichols
Joseph Mills
William McCoy, Jr.
Kelly Arnold
Linda M. Benson
William McNeel
Phoebe F. Heishman
Roy G. Owens
Bill Fouch
Robert A. Byrne
William T. and Marg Hood
Jim Comstock
John C. Ailes
Jim McGoldrick
D. Boyd Dotson
James H. Jackson
Sally M. Susman
Ed Given
Edna Dotson
Virginia Nicholson
Robert S. Earl
Associated Press

A Partial List of Photographers

Doug Booth
Dave Owen
Greg Hunter
Nancy Abrams
Dale Sparks
Ron Rittenhouse
Bob Teets
Bob Fike
John Warner
Micky Murphy
Mike Myer
John Douglas
Linda Buzzerd
Bruce Carter
Tom Coter
Nancy Isner
Edwin Propst
Rob Tetrick
Suzanne Parks
Schauer
L. Bernard
Drew Moody
Tim Grobe
Bob Thayer
David Kennedy
Rick Runion
Terry
Tom Schley
Gene Puskar
Steven Wayne Rotsch
Doug Chadwick
Rick Barbero

Many photos in this book were not individually credited by the newspaper in which they appeared.

Page 2—THE MOOREFIELD (WV) EXAMINER, Saturday, November 16, 1985

In A Bag Of Clothes From Martinsburg, WV

Hi! my name is Candy Ann MADISon and I'm giving you the cloths and shoes and I hope they can do you good becavse I really fell bad about people house getting washed away and I will alway help you when you need help

your friend Candy
ann MaDISon

Estimated State Damages

(From the Governor's Press Office 11/29/85)

Dead:	**38**
Missing:	**10**
Number of People Homeless:	**2,587**
Total Dollar Estimate:	**$500 million**
Private Homes:	
Destroyed	**4,389**
Condemned	**762**
Major Damage	**3,033**
Minor Damage	**1,800**
Private Businesses:	
Destroyed	**131**
Condemned	**23**
Major Damage	**474**
Minor Damage	**315**
Public Buildings:	
Destroyed	**23**
Condemned	**10**
Major Damage	**70**
Minor Damage	**9**
Bridges:	
Number Affected	**103**
Number Closed	**48**
Number Needing Repair	**55**
Highways:	
Number Affected	**429**
Number Now Open	**375**
Number Now Closed	**54**
Schools:	
Number Affected	**15**

Note

The author of the prologue and epilogue relied on speculation about certain activities in the telling of these stories in regard to certain people. This speculation was based on interviews with friends, family members and others associated with those persons.

Cover photo by Nancy Abrams, Dominion Post

ALBRIGHT—The raging Cheat River as it crested at 9 a.m. Nov. 5. (Courtesy of Dominion Post)

Prologue: Birth of a Disaster

By Bob Teets

Sunday, October 6, 1985

Coral Gables, Florida

The computers in the National Hurricane Center along South Dixie Highway in Coral Gables, Florida, were busy tracking a strong storm brewing off the coast of Haiti.

Meteorologists huddled around print-out sheets and monitors trying to guess where the storm would head. A computer simulation indicated it would follow a generally north/northeasterly course.

If the storm were to intensify, and indications were that it would, the area of low pressure might grow into the season's fifth Atlantic hurricane.

Her name would be Isabel. She could be a big one.

For now, though, she was still just another of the one hundred or so "seedlings," potential tropical storms or hurricanes, tracked by the Center in a typical year.

Normally, of the 100, about 10 reach tropical storm status, and of those, maybe six develop into true hurricanes.

The record belonged to the year 1916, when a total of six 'canes and two tropical storms made landfall on the U.S. mainland.

Still, if Isabel were to break the odds and develop into a hurricane which would then hit the mainland, the 1985 season could be chalked up as one of the worst years on record for damage. She would enter an elite, if infamous, clan of four previous killer storms (beginning with Bob on July 21; Danny, August 12; Elena, August 28, and Gloria, September 16) which had laid waste to an estimated $2½ billion worth of property and which had laid at least 18 Americans in their graves.

But despite the concern throughout the Center for the fledgling Isabel's progress, not all attention this day was put toward her.

A simple notation made by another observer in the Center denoted the birth of yet another seedling some 3,500 miles away just off the coast of the Sahara Desert in Africa.

It was headed out to sea on a familiar course toward the Americas.

It would prove to have been the birth of a West Virginia disaster.

★★★★★

Albright, West Virginia

Jack Wotring, as usual, hadn't slept well, so he lifted himself from his "bed," actually his favorite easy chair. He ambled across the thickly carpeted floor of the living room to look outside.

He mused first on his reflection in the window: the outline revealed that the tough, former World War II infantry man who later became a master plumber and builder had gone slightly soft after retirement when perhaps 200 pounds had piled up on a hefty five foot, eight inch frame.

But the image also showed the familiar swept back hair cut and wide, dark, expectant eyes full of humor. He thought the oblong face staring back at him had been weathered some by 61 years of active living, but they had been good years—proud years of hard work and solid accomplishment.

So what if the blasted old heart had forced retirement, it wasn't about to force him into a rocking chair.

No sir.

Keep busy. Keep moving, keep moving.

He looked outside.

The early morning sun sent its first rays charging headlong into the bright waters of the Cheat River 200 feet away and promised to break the night chill which had crept over Briery Mountain and down into Albright.

Jack saw no one on the streets of town at this early hour on Sun-

ST. GEORGE—The killing waters of November thrust one of its victims into the debris-packed supports of this Tucker County bridge. (Photo by John Warner)

day morning, so, alone, he silently watched the birth of a new day.

★ ★ ★ ★ ★

Monday, October 14

Hurricane Center
Coral Gables

Satellites high above the earth screeched out information to the receiving station at Wallops Island, Virginia, where it was then relayed directly to the whining machines in the computer room at Coral Gables.

A few eyes scanned a monitor as a cloud formation jerked in steps across the screen. Still pictures taken by a weather satellite and then projected quickly one right after the other on the monitor gave the appearance of movement to the formation.

The image was that of Tropical Storm Isabel, which had delivered a soft punch to Georgia and was now spinning harmlessly away onto the Atlantic.

But another seedling was causing some concern as a sequence of pictures on a separate monitor showed it splitting in two.

The first half was flitting off toward obscurity in the Pacific Ocean.

Mark Zimmer, staff meteorologist, noted that the second half joined a large mass of unstable air near the Gulf of Mexico.

Zimmer knew if it were to form into a tropical storm, it was to have the next masculine name on the Atlantic list, because Isabel had already taken the feminine one.

The names were drawn from those contained in books which list possible names for babies.

This baby would be known as "Juan."

★ ★ ★ ★ ★

Albright

Jack Wotring noticed the Cheat was low. The river was thirsty, he thought, just like his garden and his prized apple trees, the latter of which was just this year really starting to produce. Already, several friends from Kingwood and his neighbors in Albright had come over for a sack-full of "Jack's apples."

This morning, his bad leg, probably the latest target of that cursed diabetes, ached just enough to make him want to do something to take his mind off it.

Keep moving! Keep moving!

He switched on the television set, then headed into the bathroom to clean up and shave.

By the time he finished, he could smell the buckwheat cakes and sausage that Maxine—his wife of 39 years—had frying on the griddle in the kitchen.

As they ate, Jack and Maxine paid little attention to a newscaster on the television speaking of a tropical storm named Isabel which had harmlessly swiped the coast of Georgia on Friday after flash floods caused by the storm's rains had killed 100 people in Puerto Rico earlier in the week.

Friday, October 26

Preston Country Club near Kingwood, West Virginia

Lynn Taylor, 89, got out of bed around 7 a.m. and, as was his habit, put on his old, weathered, black and red plaid Mackinaw jacket for a quick walk and a breath of fresh air outside.

He was a distinguished-looking man, the professor type, with white hair, blue-on-blue eyes, and the clean, athletic, sun-tanned features of his prime which he'd carried into these "later years," a term he would refuse to acknowledge.

His eyes did betray a streak of stubbornness, but a lifetime of fruitful living had repaid him with a good sense of humor. That shone there, too.

As he walked out onto the porch this morning, he breathed deeply a crisp West Virginia air that, on December 13, would be much warmer when he landed in Hawaii to spend the winter with his sister, Dolly.

For now, though, he intended to follow his morning routine, and paused just long enough to appreciate the exhilarating scenery.

His cottage was separated from the broad Cheat River by a narrow road leading past the country club to Camp Dawson (the county's national guard armory), a half-mile or so up the valley. He

MARLINTON—Organ caked with mud drying out in front of St. John's Episcopal Church in Marlinton (Courtesy of Elkins Inter-Mountain)

scanned the familiar golf course and the exquisitely manicured lawns of the homes off on the far side of the club house. He felt as he did every morning—that it all belonged on a picture post card.

Finally, he stepped out onto the rock walkway, then onto the drive. He rolled out and climbed onto his old bicycle, and headed down the river road, humming a song (slightly out of tune) above the noisy waters of the Cheat.

"Sunset and evening star . . ."

★ ★ ★ ★ ★

Somewhere in the Gulf of Mexico

A thousand miles to the south, a storm flung its gale force winds into the warm Gulf waters.

At the Hurricane Center in Coral Gables, a name had just been assigned.

The storm was gaining momentum, its voracious appetite feeding from the energy released by the heated tropical water, its temperament worsening with each mile it churned.

And deep in the interior of the angry, swirling clouds, a calm spot was forming.

Soon, the monster would have an eye.

Soon, Hurricane Juan would see.

★ ★ ★ ★ ★

October 31, 1985

For six days, Juan had abused the Gulf coastline from Port Arthur, Texas, to points east and south, spinning ruthlessly out of control.

And now, on Halloween, the wicked one lashed out. 85 mile-per-hour winds forced sea waves to bash in a house and rip up the shoreline of Captiva Island, Florida.

The local newspaper, *Island Reporter,* would later write that the "unwelcomed visitor from the tropics . . . named Juan, made up with tenacity what he lacked in strength."

Total cost of the southern states' battering: $1½ billion in damages; 12 dead.

In Coral Gables, Juan was about to be written off now that he had made landfall. He would lose energy and track off to the north as a rain storm.

But Juan would not die.

For him, Halloween was just beginning.

Saturday, November 2

Country Club

Lynn Taylor thought the rain would never stop.

The television was constantly showing pictures of the hurricane which had slammed on shore down south and which was now believed to be causing the heavy rains in West Virginia.

Well, no matter, he had plenty to do and plenty of time to do it.

He methodically prepared breakfast. It was almost a ritual with him.

First the coffee. He liked his own the best, much better than Dolly's.

He brought out the egg cooker, two eggs, a bowl, a large and a small plate, a sharp knife, a small pitcher of pure cream (oh, that was a must!), a box of cereal and a grapefruit.

RIVERTON—An aerial view of the damage done by the North Fork at Riverton.

(Courtesy of Clarksburg Exponent Telegram)

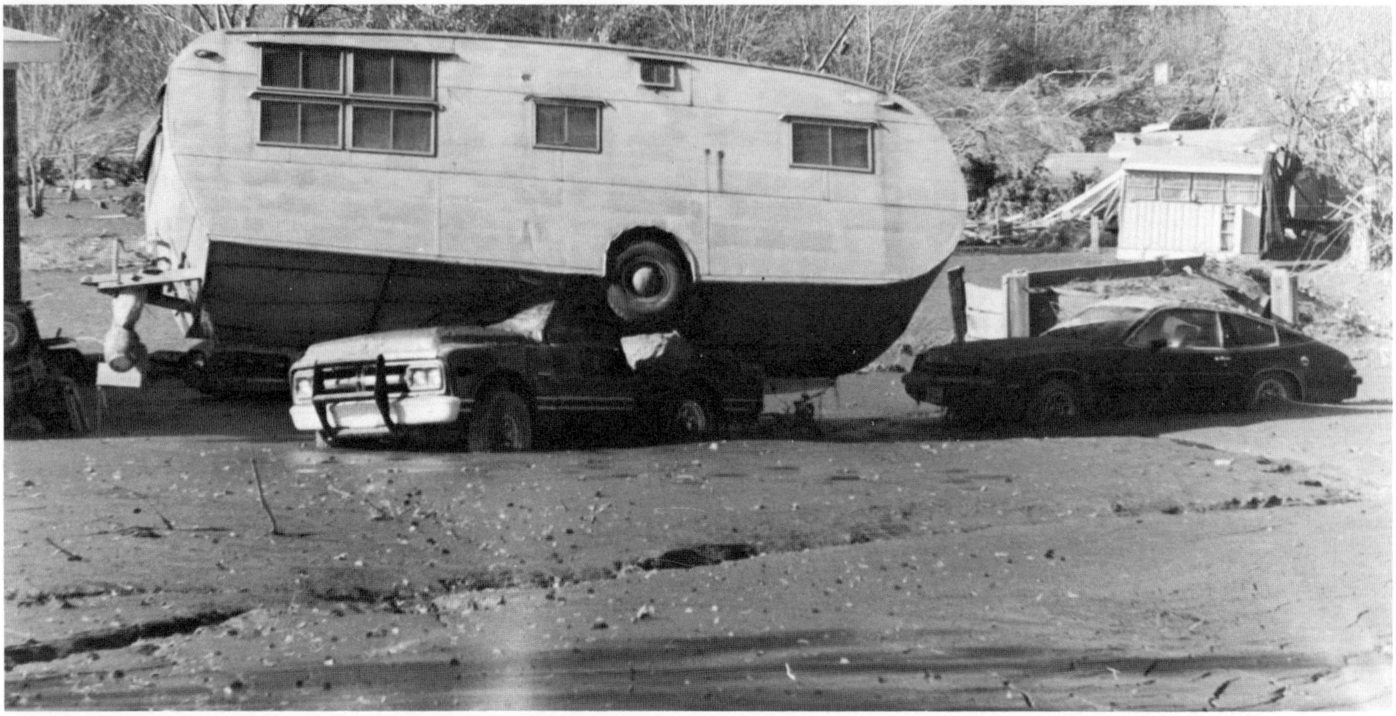

ROMNEY—Floodwaters from the South Branch of the Potomac River carried this trailer to its final resting place, on top of a pickup truck. (Courtesy of Hampshire Review)

Next, the dry cereal (unsweetened) was poured in the bowl and placed on the neatly arranged table where a small vase held a clutch of slightly wilting wild flowers he'd picked the day before on his regular bicycle ride, taken between cloudbursts.

In a few minutes, the meal came together nicely, everything timed to the second, and he placed it on the table.

Then he sat and bowed his head.

Never would he touch a meal without first praying.

Religion was important to Lynn Taylor. He liked the Episcopalian traditions and rituals. The faith was his rock, his island.

★★★★★

St. Marys, West Virginia

Jack and Maxine Wotring had driven down for the weekend to see some of the family and to allow Jack a chance to spend some special time with a grandson.

So Jack and the boy were out tromping around in the woods on a farm not far from town, and Maxine was at the house tasting the daughter-in-law's hot pepper relish for the first time in her life. She loved the recipe and vowed to fix some of her own when they arrived back home tomorrow.

★★★★★

Sunday, November 3

"Ring that bell!"

And the three strands of sleigh bells filled the cottage with the sounds of Christmas past, deadened only slightly by the sound of yet another cloudburst outside.

Lynn loved his collection of bells, from the big sonorous one outside on the post in the yard—which he would occasionally club first thing in the morning to tease his house guests, and half the rest of the valley, for that matter—down to the smallest crystal one on the mantel that gave off a single, pure and petite "tingle" when it was gingerly handled.

There was no one else in the cottage this morning, but he shook the sleigh bells anyway. Why not? It was Sunday. Bells especially belonged to Sunday!

He'd already eaten breakfast, and there was still plenty of time before church.

"Humph!"

For some reason, a scene from a church in Hawaii flashed through his mind.

He'd gone there with his sister Dolly.

Before he knew what was happening, a choirboy had placed a lei around Lynn's neck.

And right out loud, with a red face looking even more stark against his white hair, he'd protested, "I don't like those damn things!"

That was the last trip to that church, by gosh. Not enough ritual, not enough reverence!

★★★★★

St. Marys

Jack and Maxine Wotring pulled out of the driveway in mid-afternoon.

ALBRIGHT—Jack and Maxine Wotring (Courtesy of Preston County News)

PETERSBURG—A new but never completed rail spur along the Johnson Run Road into the Petersburg Industrial Park is left twisted from the force of the water. (Courtesy of Inter-Mountain)

On the way back to Albright, the conversation drifted from hot pepper relish to the growth of the grandchildren to the delicious breakfast which Jack and his grandson had helped provide . . . creamed squirrel and buckwheat cakes.

★★★★★

Country Club

Lynn called Dolly as soon as he returned from church.

"Well, I called the airline," she said, "to confirm your flight for the first. I mailed your ticket yesterday."

"I'll be ready," he said expectantly, looking out at the miserable day where the steady rain continued to beat down what few leaves remained on the trees in the yard.

Dolly sensed a loneliness then in her brother's voice and knew he needed bolstering. "I think it was a good idea to move up the date."

"Definitely," he said.

As they continued talking, there was a persistent memory pricking Dolly's mind. No matter how often he tried, Lynn would always meet her or part from her at the airport with the glistening of tears on his cheeks. "He may attempt his gruffness on the rest of the world," she thought, "but I know the real man, the kind, loving, charitable man that he truly is."

She usually said "Aloha" to him when the call ended, but for some reason, as this conversation wound down, she said, "Lynn, I love you," and waited for a response.

"Huh? O-Oh, yeah, yeah," came the gruff voice, "Me too."

★★★★★

Albright

Jack and Maxine returned to Albright about 6:30 p.m.

After dinner, Jack settled down to watch the TV wrestling matches from Atlanta.

An announcer broke in once to bring the news of the progress of recovery efforts along the Gulf Coast following the arrival of Hurricane Juan.

But already Jack's mind was thinking about what he would be doing tomorrow.

Keep going! Move on, move on!

He decided he'd probably go over to Deep Creek Lake and check out his fishing boat moored there. Blasted rain may have sunk 'er by now!

Maxine laid out some hot peppers she'd picked from the garden, and busily prepared a list of things she would need to buy at the

PHILIPPI—The courthouse, too, was damaged by floodwaters. (Courtesy of The Democrat)

store tomorrow. Yep, Monday would be a hot pepper of a day in the Wotring house!

★★★★★

Monday, November 4

Country Club

Lynn had slept in this morning, and upon arising, had grumbled at the fog and rain for hanging like a curse over the valley.

His mood, improved, however, after breakfast, so he threw on the Mackinaw and headed out to his workshop to putter around some and to polish some silver. He figured he may as well buff up some of the brass bells, too.

"Ring those bells," he shouted, running through the rain to the workshop. "Ring those bells!"

★★★★★

Albright

After dinner, Jack sat down in front of the TV set.

Maxine was too tired to wash the dishes because she'd fooled around all day with the pepper relish and other jobs, so she decided to take a nap and do the dishes later.

Jack was fairly content for the moment, having decided that he had enough coal—8½ tons—stocked in for the winter. Better to be safe than cold and sorry. "Now, the younger people," he thought, "yeah, sure, they just assume everything will be taken care of. But we older folks try to be prepared for tomorrow. Hell, I like to be at least as smart as a chipmunk! Lay a little bit away for winter time."

The thought of winter triggered a brief memory of the cold waters off Omaha Beach during the landing. The mortar shells. The death.

"Keep moving!" the officers commanded. "Keep moving!"

★★★★★

Across West Virginia

Meteorologists were to debate the importance that Hurricane Juan would play in what was about to happen in West Virginia. Some would contend Juan had already passed from the area and contributed little. Others would say the remnants of the killer hurricane left enough moisture in the area to break the proverbial camel's back when another storm system brought in its share of rain. Still others would say that Juan's swan dive created a sort of vacuum which pulled moist air in from the Atlantic and joined it with a storm front which stalled over the mountains and created the worst flooding in the state's history.

But for Lynn Taylor and Jack Wotring and the thousands of other West Virginians whom their stories represent, the debate about Juan would prove meaningless.

What mattered most during those fearful hours of November 4 and 5 was the knocking at the door.

Death had come calling for some 40 people. And thousands wondered: "Is it for me?"

★★★★★

(For the conclusion of this story, please turn to Page 111)

PETERSBURG—An ARCO service station became the target of fast-moving logs in the torrent. (Courtesy of Piedmont Herald)

MOOREFIELD—Moorefield school buildings and equipment suffered an estimated $1 million worth of damages.

(Courtesy of The Moorefield Examiner)

PETERSBURG—A mosaic of destruction surrounds battered buildings throughout the flood-ravaged area.

(Courtesy of Parkersburg News)

Margaret Painter swept away by flood waters at Petersburg; husband Bill rescued after 13 hours clinging to debris and trees

Nov. 8
Inter-Mountain
Elkins
By Eldora M. Nuzum

Editor's Note: Young Elkins residents Bill Painter and his wife, Margaret, were returning home from a Marathon race in Washington, D.C., when they were trapped in the flood near Petersburg on Monday. Painter's wife is still missing and his home in Crystal Springs is severely damaged by floodwater. Painter recalled his 13-hour nightmare as he talked to Editor Eldora M. Nuzum:

"We were about ten miles outside of Petersburg going toward Mouth of Seneca. There was about six inches of water crossing the road. We were able to drive through it, but then we came to a point where we couldn't cross the road, so we turned around and tried to go back the same way we came, but the water had risen to where we couldn't cross it.

"We stopped at a campground at Jordan Run. There were two other trucks with people and two hunters—nine all together. Margaret was the only woman.

"We had all parked on a knoll at the campground and it was becoming surrounded by water. It was three or four feet deep and 100 feet across at one place and we tried to ford it, but the water was too swift to cross.

"We all decided it would be safer to stay where we were. The water kept rising and about 5:30 p.m. my wife and I crossed the water to a two-story house at the campground. We waded in the water to get to it. We broke out windows and climbed inside. We went up to the second floor and at about 7 p.m. the other people who hadn't come across, tried to make it. The water had risen six to seven feet. They had a small boat and five to six people were clinging to the outside. Two older men were inside the boat. They got within five feet of the house and the current swept them away. One or two of them got back to where they originally were and climbed on top of some mobile homes. That was about 7:30 p.m.

"Ten or 15 mobile homes and trailers were all washed away.

"At about 8 p.m. the water came to the second story window. It had risen 15 feet in three or four hours. It knocked the house off the foundation. It broke up. As it was breaking up, the roof split in two and my wife and I climbed out onto the section of the roof which measured about eight feet by eight feet.

"We floated down stream about three miles on a section of the roof, then it overturned and I lost hold of her. The current was so swift that it kept me under for about 20 seconds," Painter said.

"I was floating and I grabbed a bathtub or a refrigerator and hung on to it for awhile, but I lost it. I grabbed hold of a big tree that was floating by. The tree became lodged in some other trees and I climbed up into one of the trees.

"I was in it for about one hour when the water knocked it away and then I started swimming. I could see more trees 300 or 400 feet away. I was swimming perpendicular to the current and finally reached calm water and found another Sycamore tree and climbed up in it. That was about 2 a.m.

"I started yelling and somebody heard me and yelled back. Red lights like an emergency squad vehicle came about 15 minutes later. I yelled back and forth for a couple of hours and the person said a helicopter was on its way.

"I was about 400 to 500 feet away from them. The water was 1,000 to 2,000 feet wide in places and it had calmed down. I knew I had made it then.

"By 4 a.m. the water started going down and the current started rippling over the debris. There was too much noise and I lost contact with the people until it was about daybreak and I could see the red light.

"At 6:30 a.m. or 7 a.m., I climbed out of the tree. The water had left enough debris that I could stand on. The water was still about six feet deep, but I was able to get out of the tree and stand. At 8 a.m. the helicopter came. It was a Maryland State Police helicopter.

"They dropped me a sling and I climbed out on the debris and got into it and they took me to the ambulance.

"I was right across from the Dettinbourn Trucking Co. which is a few miles from Petersburg. The helicopter let me down there.

"They said I was in the river for about six miles.

"They treated me for hypothermia at the Grant County Hospital. I stabilized pretty quick and I asked them to find me a ride back to Elkins.

"The Department of Natural Resources helicopter was flying back to refuel and they got me home Tuesday afternoon.

"Of all the persons in the other party who were stranded with us, the rescuers have only recovered one of the bodies."

A second disaster for Painter occurred when he returned home on Tuesday to find his Crystal Springs home had been flooded.

"There was 18 inches of water inside the house and three feet of water in the garage. Friends are helping to clean it up."

Painter's wife is the former Margaret Lilly who is originally from the community of Daniels near Beckley. She is 29 years old.

She has been employed for three years as a pharmacist at Memorial General Hospital.

They are the parents of a 21 month old son, Jonathan, who had been staying with Painter's mother, Frankie Painter, over the weekend. She resides at Riverview Addition in Elkins.

Bill Painter works for the Leadsville Public Service District which is helping to install the new sewer system in Elkins.

He had formerly worked for Parker Leonard at the Hillside Nursey in Elkins.

PARSONS—Vittie and Thelma Lipscomb stand on the front step where their home once stood. The house, in the background, was moved approximately 50 yards off it's foundation. The Lipscombs were in the house when it moved. (Photo by John Warner)

NATIONAL WEATHER SERVICE RIVER STATISTICS: 11/07/85

TYGART VALLEY RIVER
- ★ Dailey - crested at 16.5', flood stage 14'
  2 a.m. Nov. 5
- ★ Belington - crested at 24', flood stage 14'
  4 p.m. Nov. 5
- ★ Philippi - crested at 30.5', flood stage 17'
  10 a.m. Nov. 5

OHIO RIVER
- ★ Wellsburg - crested at 33', flood stage 30'
  7 a.m. Nov. 7

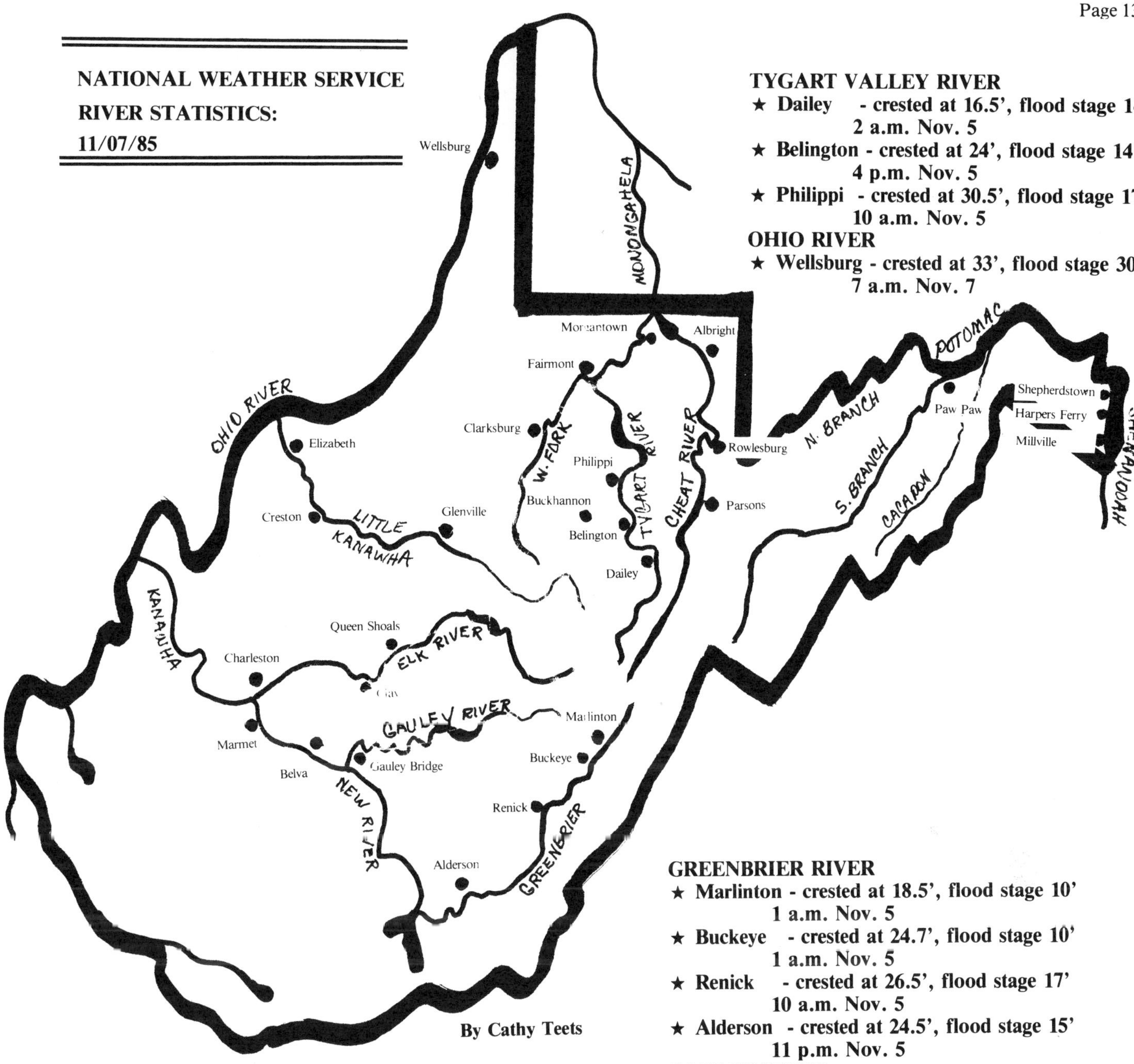

By Cathy Teets

SHENANDOAH RIVER
- ★ Millville - near crest at 31', flood stage 13'
  Nov. 6

WEST FORK RIVER
- ★ Weston - crested at 24.3', flood stage 17'
  7 a.m. Nov. 5
- ★ Clarksburg - crested at 18.5', flood stage 7'
  8 a.m. Nov. 5

CHEAT RIVER
- ★ Parsons - crested at 21', flood stage 13'
  3 a.m. Nov. 5
- ★ Rowlesburg - crested at 17.5', flood stage 12'
  6 a.m. Nov. 5

POTOMAC RIVER
- ★ Paw Paw - crested at 54', flood stage 25'
  late Nov. 5
- ★ Shepherdstown - crested at 33', flood stage 15'
  4 p.m. Nov. 6
- ★ Harpers Ferry - predicted crest at 34', flood stage 18'
  10 p.m. Nov. 6

GREENBRIER RIVER
- ★ Marlinton - crested at 18.5', flood stage 10'
  1 a.m. Nov. 5
- ★ Buckeye - crested at 24.7', flood stage 10'
  1 a.m. Nov. 5
- ★ Renick - crested at 26.5', flood stage 17'
  10 a.m. Nov. 5
- ★ Alderson - crested at 24.5', flood stage 15'
  11 p.m. Nov. 5

GAULEY RIVER
- ★ Gauley Bridge - crested at 23.5', flood stage 23'
  Nov. 5
- ★ Belva - crested at 11.6', flood stage 20'
  2 a.m. Nov. 5

KANAWHA RIVER
- ★ Marmet - crested at 29.6', flood stage 37'
  6 a.m. Nov. 6
- ★ Charleston - crested at 22.5', flood stage 30'
  6 a.m. Nov. 6

ELK RIVER
- ★ Clay - crested at 15.5', flood stage 18'
  5 a.m. Nov. 5
- ★ Queen Shoals - crested at 14.6', flood stage 19'
  6 p.m. Nov. 7

LITTLE KANAWHA RIVER
- ★ Glenville - crested at 36.5', flood stage 23'
  7 a.m. Nov. 5
- ★ Creston - crested at 29', flood stage 20'
  1 a.m. Nov. 6
- ★ Elizabeth - crested at 34.5', flood stage 36'
  4 a.m. Nov. 6

Residents seek refuge from Little Kanawha

Nov. 6
Parkersburg Sentinel
Parkersburg
By Dave McMillion

As water from the Little Kanawha River filled communities like Newark, Creston and Big Bend Tuesday, residents worked hard to protect their houses and valuables from the oncoming flood.

People living in higher elevations in these areas of Wirt and Calhoun counties donated space in their houses for flood-stricken residents to store personal and perishable items. Others moved from house to house in small boats rescuing friends and relatives from high waters.

Many residents could be seen measuring the water as it came closer to their houses, hoping it would begin receding.

At about 5 p.m. Tuesday, Mary Reno of Creston finally decided to move her furniture and other personal items to the second floor of her home as the Little Kanawha River edged closer to her house. "I've lived through floods on other rivers such as the Ohio and I've learned it's better to be safe than sorry."

Chip Shaver, a conservation officer with the Department of Natural Resources, plowed through the small communities along the Little Kanawha River by boat making sure people who had been isolated by high waters were safe.

Members of the Wirt County Volunteer Fire Department assisted Shaver. "We're just seeing if people want to come out or see if they need anything," said Walter Hupp, member of the Wirt County Fire Department.

David Miller of Creston kept a close eye on the high water. "I've been watching this water come up all day," said Miller.

Although Miller was expecting the high water to fill the first level of his house, he said it would not be as bad as the flood of 1967. "The water was up over the banister (on the front porch of his house) in 1967," he said.

The Little Kanawha River was expected to crest in Creston at 34 feet, 14 feet above flood stage, Tuesday night. Miller said the flood of 1967 reached 37 feet, 17 feet above flood stage.

Residents of Big Bend, particularly near Stocking Run, were flooded by the Little Kanawha at about 3:30 p.m. Tuesday.

Water in the area rose faster than expected. So fast that Joe Kane, a veterinarian, almost forgot about his six bulls in a barn near the Little Kanawha.

Oren Parks of Big Bend said Kane had to lead the bulls out of waist-high water and through heavy brush. "They liked to never got them out of there," Parks said.

Stratton is tops

Nov. 18
The Gulf Times
Beckley
By Sally Matz Susman

The student body of Stratton Junior High School has come up with a great idea and I think we all should help them along.

Teachers at the school thought of the project first and passed it along to the Student Council who in turn decided it should be a school project. The students were only too happy to join in, according to instructor Jess Farley.

The youngsters want to help the students at Marlinton Middle School in Pocahontas County. As you know, Pocahontas County was hit hard by the recent flood and the Marlinton Student Body is the most needy in that area.

In order to aid the flood victims, Stratton students have adopted Marlinton Middle School students and will spend the next week raising money to buy items that are needed.

Thomas Sanders, principal of Marlinton, told Farley what was needed most: toothpaste, tooth brushes, combs, hair brushes, bath soap, deodorant, shampoo, hand lotion, notebooks, pencils, pens, candy, and notebook paper.

Farley, who is also sponsor of Stratton Student Council, said each student is to contribute at least $1. The money will be given to Sanders who will distribute it among the most needy of his students.

In addition, Stratton art students have designed gift bags and the English Department has composed letters for each individual Marlinton student. The bags will be filled with combs, pens and pencils, etc. The letters will also be included.

That's not all! Stratton students are adopting the Marlinton students for the entire year. Money will be raised to give the needy students individual gifts at Christmas time. In addition, the faculty and students of Stratton hope to raise enough money to buy the 25 most needy students new clothing outfits.

MOOREFIELD—Staff and volunteers at Moorefield Elementary School have almost completed their cleanup operation. (Courtesy of The Moorefield Examiner)

PARSONS—Trees were uprooted and the faces of homes were sheared away by the force of floodwaters along Pennsylvania Avenue in Parsons. (Courtesy of Elkins Inter-Mountain)

A newspaper editor's diary

Nov. 13
The Parsons Advocate
Parsons
By Mariwyn McClain Smith

The following is not an attempt to tell what happened to Parsons, Tucker County, West Virginia on the night of Monday, Nov. 4, 1985. Of necessity, I can relate for you only what happened to me and to those whom I care about most dearly. I'd been released from Tucker County Hospital just before noon. An abscessed tooth my dentist had been unable to get under control, in spite of a week of medication and bedrest, had sent me to the hospital with chills, a fever, and dehydrated.

Monday, George brought Faith Anne (our 16-year-old) home with him at 6 and as soon as we had dinner, he headed back to the shop. It had rained for days, and George was worried about possible flooding, but our home is the highest on "the island" (Pennsylvania Avenue) and we'd joked for years that if water ever got into our home it would be time to build an ark.

George left at 7 and in less than an hour we were stranded. I'm not certain anyone really truly knows what happened in that hour. Where that wall of water came from may forever stay a mystery. New riverbeds were formed. We watched neighbors in three other nearby houses and knew we weren't the only ones left alive. Perhaps that's what helped maintain our sanity. However, we did take food and water upstairs. We did save our photograph albums. Unbelievable as it now seems, we did sleep through the night—even as the porch and garage were swept away.

Reconstructing, now, a few days later, I can write what others have told me. But, mostly, I can tell you how it occurred to me and my family; to our home and our two businesses—the newspaper and the printing company.

It is not a story I would ever have chosen to write, but it is history.

MONDAY NIGHT

7 p.m. George left to check the office.

("Things that go bump in the night")

About 7:45 Faith Anne said front yard was covered; Roys' truck floated into Plumleys' house; two porch posts loose.

Call neighbors; fence leaving, one rail at a time.

Cars and trucks and utility house floating down Avenue.

Gathered food, medicine, pocketbook, coats, and went upstairs.

9:47 stood upstairs leaning over the bannister, watching as water coming in under the front door turned the carpet from off-white to mud brown. It was pouring in through the furnace ducts.

Familiar voices over the scanner from all over the county; comforting, so good to know they're willing to leave their homes to care for others.

10:00 Sounds like a column gave way from the front porch but the roof seems intact.

I'm grateful I'm not alone.

10:05 Terrible grating noise; in our bedroom; sounds like the porch pulled away.

10:10 Water up to third step—morbid, can't look at what it's doing to my home, but can't bear not to go from room to room—F.A. screamed "A house! Mom! a house." What appeared to be the entire roof of a large garage went in front of us, down the drive and behind Plumleys'.

Things downstairs are clinking and a big log is going thru the doorway.

10:40 first tears—my piano; water up to fifth step. Front door opens and closes, opens and closes, shaking the house.

So many things we could have carried upstairs.

My God! How foolish I was!

10:45 Fikes' house has been shoved off its foundation, looks like 45° angle. We think they're in it. (We learned the next day that Mr. and Mrs. Denver Fike and her 99-year-old father were there. They survived!)

10:55 Electric off—at least we'll now stop watching the clock.

11:08 Transformers blew near the old gym, half a block away.

We see flickering lights at Fikes'. (They were candles.)

TUESDAY MORNING

6:15 I don't know how, but we both slept thru the night. Might as well get up.

The water's receded and is gone from the living room, but there are three feet of mud, a large tree stump and a million sticks in there.

From the front door I could see all our front porch and the pillars are gone—the roof is still there! A pickup truck is resting against the house. Our garage, car and truck are gone. Can't close the front door because of the mud.

(Please, God, let George and Ken and Ruth be okay.)

By 7 o'clock we know the entire front of the Wilson apartments is gone (looks like a doll house with one end open.)

Plumleys have lost their patio, utility building, car and truck. Ed and Lola are clearing mud out of their house. Our car is over the hill behind their house.

The Black Fork is level with our back yard.

8:30 The water is receding. Bonners' house is off its foundation. Pauline Harris' house is surrounded by trees and appears badly damaged.

There's no sign of life at Wimers'. (We learned later that they had left and are okay.)

An immense hole, several yards wide and several feet deep has been washed out between our house and the Plumleys'. A new riverbed has been cut on the other side of Wimers'.

We threw stuff out the livingroom door for a while, but it seems useless—getting no where—at least the upstairs is clean, dry and reasonably warm. We have some water, food, and each other.

We went back to bed and slept an hour. A man shouting "Mrs. Smith" awoke us at 9:30. He's from the DNR, apparently came in thru a den window (they're both broken.) George sent him—he's okay, thank God! He was riding a kayak and said if we were in danger, they'd send a boat for us. We told him we were okay, that we'd wait until the water went down, then walk.

10:30 The hole between our house and Plumleys' is widening.

1:45 We're at Ken and Frank's house. Neil Lamson and Mutt Carr came in thru our back door about 11 o'clock and walked us out between our house and Wimers', across what was left of Rt. 219 (chunks of macadem the size of a VW laying around), and behind Brocks' house to the railroad tracks. We walked the tracks to behind Swartzs' garage, then to the Big T where we thought George was waiting.

Faith Anne brought her camera and Tigger, zipped up in a duffel bag because she was afraid they wouldn't want her to keep him.

We were there outside of the Big T about 10 minutes when Ken drove by. He brought us here—so good to see our men! We were soaked to our knees and glad to be in a warm house. Ruth's Aunt Jenny and family were there—12 of us, in all.

Every store on First Street is wiped out!

The railroad bridge, parallel with the downtown bridge, has several feet of garbage piled on it.

The war memorial on the courthouse square fell over and is in 4 or 5 pieces.

2:55 The electricity is on!!

George and Faith Anne left about 2 o'clock to see if they could get back to the house.

Ken left about 3:15 to see if he could get to our house. Frank and Ken got back about 2:30 with water, sodas. They had stopped at the office.

George is distraught! He's ready to file bankruptcy. Water was five feet deep in our offices (my bookkeeping was at home because I'd been sick). Everything in the shop was under water and he's afraid the presses will never run again.

I wish I knew about Barby.

3:30 I'm tired. It's all catching up with me.

The porch roof (at our house) fell off while George and Faith Anne were there. There's no sign of structural damage—if we can just get the mud out.

WEDNESDAY

We keep hearing stories, no way of knowing what's true.

Everyone hugs everyone.

Everything is beginning to stink (including us)!

We came back to Ken and Frank's to eat when it got too dark to shovel. I hurt!!

Some strange things didn't get broken—glass globes for candle holders, china, lamps....

It's been less than 48 hours—it seems like weeks.

THURSDAY

Today was "recovery day." My shoulders ache so bad I could cry.

Chris, the pharmacist, went with me to the hospital to get a tube of rubbing ointment. Janie Jones rubbed by shoulders—it felt so good!

Barby is here! There's really nothing she can do, and she was safer staying in Shepherdstown, but we were glad to have her home.

Town is full of National Guardsmen, they're asking people to stay off of Pennsylvania Avenue. We did, but many, many others didn't. We can't imagine where all the truckloads of gravel are coming from—dozens of them—filling in around and under the Sluice Bridge and the Black Fork Bridge. Once they're repaired we won't be so isolated; Rt. 219 to Thomas will be open.

We slept this afternoon; felt guilty because we weren't "doing" anything, but really just couldn't work up the energy. Ken and George stuck it out at the shop all day. They're anxious to see if the equipment can be cleaned, oiled and repaired.

It was raining hard and the wind blowing when we came down—like rubbing salt in an open wound.

FRIDAY

We had a good night's sleep, in a real bed.

Darkness was almost welcome. You just eventually wear out of steam.

All our crew went over to the fire hall to plan what needs to be done tomorrow. The Salvation Army crews are set up there.

A group of Civil Air Patrol cadets came from Charleston early this week. They are teenagers and apparently put in a very hard day of work.

Two or three people have described Parsons as looking like "a town under seige" and one added, "with no one shooting."

I understand three young men were arrested this morning for stealing 4-wheelers. We're told they're to be charged with grand larceny.

An example of how fast the flood occurred: Frank Ours drove up Pennsylvania Avenue Monday about 7 p.m. to get milk at Jim's All-Star Market. About 15 minutes later he came back and his car began to float. He deserted his car and discovered the water was up to his hips. The Black Fork River apparently rose 3 or 4 feet in 15 minutes!

It could be worse.

SATURDAY

It was 33° when we got up this morning.

Back to our home.

Faith Anne and I went upstairs. The mud on the stairwell is nearly dry so we didn't bother taking off our filthy sneakers. We filled pillowcases with underwear, socks, a luxury for me—a bathrobe, George's filthy mud-filled hunting boots, a long extension cord, and

I was very pleased to find my good warm car coat in one piece.

We flagged down Bob Pase who drove us to Ken's; then we flagged Scott (Todd?) Bolyard who drove us down to the shop to start cleaning.

Cleaning the shop today we made terrific headway. As you're doing it, it seems hopeless! The hallway between my office and George's is about seven feet wide and 40 feet long. We had two water hoses and four people with brooms. It took us well over an hour to turn the mud to reasonably clear water. My offices have been hosed out twice.

We're told that Lambert Chevrolet lost 62 new cars and several motorcycles and/or four-wheel drive vehicles.

It's 11 o'clock and I'm ready for bed. We're tired but we're gonna make it!

Opening schools is still very much a question. Hamrick, Parsons, St. George and the high school are all being used as shelters. Most everyone there agreed, however, that the football team should be encouraged to go to the state tournaments since they have had such a good season. Elkins has offered use of their field and equipment for practice.

MONDAY

We've lived through a week of this. Now we begin discovering who we can really depend on. It's been wonderful to have the immediate help, but now begins the honest-to-God drudgery of trying to put our lives back in order.

Two weeks from now . . . two months . . . a year from now . . . we'll know how we, or if we, survived.

"If it hadn't been for those people (volunteers who knocked on the door), we probably would have went to bed and never would have made it."

—— C.W. Caniford, flood survivor

PARSONS—Ronald Lee Cosner begins the laborous task of cleaning out the mud in his grandmother's house in Parsons.

(Courtesy of Huntington Herald Dispatch)

Flood leaves destruction - four dead

Nov. 7

The Pocahontas Times

Marlinton

For years reports on flooding possibilites in Marlinton have been described in federal agency language by what is known as the "100 Year Flood." On Monday citizens in Pocahontas County found out what the effect such a flood would have on the County and the results are truly tragic—four dead, the Town of Marlinton almost totally devastated, a portion of Cass heavily damaged, three Greenbrier River bridges washed away, and other sections of the County damaged.

The heavy rains that caused the flooding in Pocahontas County also caused widespread flooding over much of West Virginia, as well as other states. Record flood levels were reported on the Greenbrier, West Fork, Cheat, Tygart, and Little Kanawha Rivers.

Four people are known dead in the County due to the flooding and authorities are not sure of the location of several others.

Leo Mace, Sr., died in his Marlinton home when he was unable to escape from the rising flood. Due to damage to the VanReenen Funeral Home the body was taken to a Lewisburg funeral home.

Three people died when the car they were traveling in was washed off of Rt. 39 about 7 p.m. Monday. Dead were two women, Ruby Skeens and Rose Mick, and a four-year-old boy, Darren Skeens. They were traveling east and the car was washed off the road above Devil's Backbone just east of Huntersville. The vehicle was found about 200 yards above the Backbone. The first body was located about 3:30 p.m. Tuesday and the others shortly thereafter. All three had been washed from the car. Ms. Mick was found about a half-mile below the car, the child about a mile below, and the body of Ms. Skeens about 1 1/2 miles from the car. The bodies were also taken to a Lewisburg funeral home.

Details on the storm that caused the widespread flooding are sketchy at this time. Rain gauges in the Marlinton area showed a total of around 7 to 7 1/2 inches of rain over the weekend and on Monday. As the rain in Marlinton didn't seem sufficient to cause such massive flooding, it is assumed there must have been heavier rains in the more northern sections of the County.

Information available to the National Weather Service also didn't predict a flood of the magnitude that occurred. For most of Monday the NWS was predicting that the water in Marlinton would crest at 14 feet, which would have been about a foot higher than the 1967 flood crest. Later the prediction was changed to an 18 foot crest, but by that time the information was too late to do anyone any good.

As best as can be figured now, the crest in the river at Marlinton was 21.8 feet. Buckeye was about 24.5 feet. After holding at the same level the water began to recede about 2:30 a.m. Tuesday. Before dark Tuesday, the river and creek were back in their normal channels.

The amount of damage in Marlinton is almost impossible to com prehend unless it is seen. All of the 60 businesses were damaged, most very extensively. About 95 percent of the over 500 homes were affected—many were totally destroyed and others badly damaged. With the exception of the parts of town above the floodplain, virtually all of Marlinton was covered. Only the Court House and houses on the east side of Tenth Avenue were untouched by water.

The water reached onto the first floor of the Pocahontas Memorial Hospital. The patients spent the night on the second flood. On Tuesday most were transported to the Greenbrier Valley Hospital at Fairlea and a few were taken to Denmar Hospital.

Pocahontas Continuous Care Center, which is supposed to be three feet above the 100 year flood level, had 18 inches of water. The residents have been moved to Denmar Hospital.

The Marlinton Elementary and Middle Schools suffered extensive damage. The new elementary building and music building, which also supposedly were built above the 100 year level, had about four feet of water. The middle school buildings and the elementary rooms in the Board building had deeper water—almost to the ceiling in the cafeteria. The building housing the gifted and craft classrooms was moved off its foundation and jammed against other buildings.

Space does not allow a business by business description of the damage except to say it was extensive in almost all cases. Not only were stocks of merchandise destroyed but buildings were damaged in many cases.

Water rose in the Marlinton Municipal Building into the first floor and covered the fire trucks and ambulance of the Marlinton Fire Department.

Burns Motor Freight suffered a large loss both to its new building, where water came into the first floor, and to trucks and other equipment.

The water carried away over half of the logs and sawn lumber at the Cramer sawmill at Stillwell. There was also damage to the mill buildings.

Ricottilli Rustic Fence at the Fairgrounds suffered in the same way, losing logs and bundles of finished fence rails as well as damage to the mill.

The Hanover Shoe plant had about 18 inches of water.

Mobile homes were particularly hard hit and there is probably not one in Marlinton on its foundation. Burns Trailer Court at Riverside was totally destroyed with not a single mobile home in place. Some washed into Burns Motor Freight and others down the river.

All residential areas were damaged with Sixth Street, lower Third Avenue, and Riverside particularly hard hit, with a number of houses washed off of their foundations. Outbuildings were spread all over.

Churches were not exempt. The Methodist and Presbyterian Churches had water completely filling the basements and the Episcopal, Nazarene, Grace Baptist, Fairview Fellowship, and Jehovah Witnesses churches had water in their sanctuaries.

Water got into the Mountain State Telephone Company building about 8 p.m. Monday and ended phone service in the 799 exchange. Electric power in the Marlinton area went off at 10:55 p.m. when the line across the river at Sixth Street was damaged.

The Durbin-Bartow area lost power Monday evening at 11:45 p.m. but it was not due to flooding. A transformer blew at the Frank substation.

A number of residents and business people spent Monday night in their homes or businesses and were not gotten out until after daylight Tuesday. There were one or two incidents where people were gotten out just in the nick of time.

People forced from their homes Monday went to be with relatives or friends or one of several places made available. These we know about include the Hillsboro School, Trinity Baptist Church, Marlinton Motor Inn, and the Municipal Building. Snowshoe is also now housing people.

Assistance for the stricken town began to arrive early Tuesday. We will make no effort in this article to list where all the help has come from. Included were National Guard units, Civil Air Patrol, additional State Police, Coast Guard units, and several fire departments. One of the white water rafting companies, Wild Water Expeditions, brought one of the boats they use on the rafting trips.

Food for both residents and rescue workers has come from many sources and we hope to be able to list as many as possible next week.

Clean-up work began in earnest on Wednesday using equipment brought in by the National Guard and others.

Outside of Marlinton, Cass was the hardest hit community. The flood waters did extensive damage on the east side of the river at Cass. The road was washed out east of the bridge and several buildings washed away.

A number of the camps along the river at Seebert and elsewhere were damaged or destroyed by the high waters. Two cabins and the campground on the river at Watoga Park were destroyed.

High water also caused problems at Durbin, Bartow, and at Green Bank.

The two old bridges across the river at Sitlington and Clover Lick and the fairly new bridge at Seebert were washed away by the water. A bridge on the Wesley Chapel road near Green Bank was also washed out.

MARLINTON—Marlinton under water early on the morning Tuesday, November 5. The water had dropped several feet by the time this photo was taken. (Courtesy of The Pocahontas Times)

MARLINTON—A house swept along the rain-swollen Greenbrier River slams into the West Virginia Route 39 bridge at Marlinton as streams in much of the state reached flood stage after several days of rain. The Pocahontas County community was among dozens in West Virginia hit by the worst flooding in a century.
(Photo by Beckley Register Herald, Associated Press Wire Photo courtesy of Huntington Herald Dispatch)

MARLINTON—Flood damage on the east side of Cass. (Courtesy of The Pocahontas Times)

MARLINTON—Cramer sawmill at Stillwell on Tuesday morning, November 5. The water had dropped several feet before the photo was taken. (Courtesy of The Pocahontas Times)

MARLINTON—A summer camp along the Greenbrier River at Seebert in Pocahontas County was swept completely off its foundation. (Courtesy of The Pocahontas Times)

MARLINTON—The floodwaters of the Greenbrier River ripped away much of this building near Buckeye on the River Road in Pocahontas County. (Courtesy of The Pocahontas Times)

MARLINTON—The Belle Fire Department amphibian vehicle was pressed into service in Pocahontas County. (Courtesy of The Pocahontas Times)

MARLINTON—Debris piled up in front of the flood-damaged homes on Third Ave. (Courtesy of The Pocahontas Times)

MARLINTON—Interior of the Clover Lick United Methodist Church. (Courtesy of The Pocahontas Times)

Victims will live with flood long after waters are gone

Nov. 7
The Herald Dispatch
Huntington
By Dave Peyton

From 2,000 feet up, the Greenbrier Valley is still beautiful. The leaves are off the trees and the cool fall weather has produced an abundant crop of bright green grass. Here and there, pine forests dot the landscape. The farmhouses on the rolling hills stand proud and white against the gray and green background provided by nature.

But look more closely, along the banks of the meandering Greenbrier River, and the scene changes. The muddy river has receded from a record flood. And what it has left can only be described as misery for those who lived on the valley floor.

The Greenbrier twists and turns like a writhing snake between Lewisburg and Hinton. Roads flank either side of the river. Nearly all the houses and mobile homes between the roads and the river were inundated. Yesterday afternoon the water had receded enough so that residents could return to make the first tentative steps at cleaning up. Multi-covered carpets, wet with water, covered lawns. Towels, sheets and other linens were draped over shrubbery to dry. But the cold, humid wind that blew through the cloud-covered valley gave little hope that anything would dry.

Homes that had been built very close to the river, to take advantage of the ordinarily spectacular view, were damaged the most. The decks that many had built over the river slumped toward the muddy water. Some had actually fallen into the Greenbrier.

But if the sight of damaged homes was bad, there was an even more ominous scene. Here and there, foundations appeared on the valley floor where, a day or two before, houses and mobile homes stood. They simply had disappeared in the raging water.

In one area just upstream from Pence Springs, it appeared an entire neighborhood of a half dozen houses or more had been swept away by the water. All that was left were foundations. A lone pickup truck was in a driveway near one of the foundations. Its occupants--a man and a woman--appeared to be wandering aimlessly among the few ruins that were left.

Below that area swept clean by the flood, the remnants of a bridge remained. Area residents said that at least two of the houses upstream crashed into the bridge, demolishing it.

Nothing remained of the bridge's superstructure or deck. Only piers on either side of the bridge and a single pier in the middle gave any indication that a bridge had stood there the day before.

The contrast between the flooded out valley floor and houses only a few hundred feet away, on higher ground, was evident, even from the air. On the ground, the contrast was even more evident. Electricity had been cut off to all the flooded homes, of course, and elsewhere power outages were scattered.

At the Greenbrier Airport near Lewisburg, for example, power was interrupted at 11:30 a.m. Tuesday and still was out yesterday after noon. A power substation near the river went out when water invaded it, officials said. The control tower was operating on emergency power. There were few other services at the airport, however. The airport restaurant was closed but the sign on the locked door said "Reserved."

Kerosene heaters were scattered throughout the airport lobby to take the chill out of the air. Well-suited men and mink-coated women, arriving in private jets and commuter airlines, stopped at the kerosene heaters to warm themselves before getting in limousines for the short trip to The Greenbrier resort hotel.

Meanwhile, a few blocks from the airport in Lewisburg, there were no power outages, no problems. It was business as usual in that gentile old town with its profoundly Southern atmosphere.

But 20 miles away in Alderson and the other towns deep in the Greenbrier Valley, the scene was totally different. The toll from the surprise flood that damaged some homes and took away others was still being assessed. Only the first meager beginnings of the cleanup efforts had begun. And with a cold damp winter soon coming, the residents had little on which to hang their hopes.

Richwood hit with 'worst flood in 35 years'

Nov. 6
Nicholas Citizens News
Summersville

Richwood, after facing drought conditions all summer, finally received the rain it needed, but unfortunately, all at once. As a result, the town on Monday faced what Mayor Hyer Sutton termed the "worst flooding in 35 years." Between the hours of 5:30 a.m. and 10 a.m. Monday, he estimated that between 3 and 5 inches of rain had fallen per hour, resulting in bridge and road closings as well as evacuations of homes in low lying areas.

Rising water forced the evacuation of six trailer homes in the lower end of Richwood. Two of these homes were actually chained down with cables to keep them from being washed away. Both the Johnstown Bridge and the Watergate Bridge had to be closed, although reports late Monday were that the Watergate Bridge had been reopened. Floodwater damage to the underpinning of the Johnstown Bridge will delay its reopening until repairs can be made.

In addition, the Jones Meat Market was threatened, and Sutton credited Nettie Fire Department, which is working alongside the Richwood FD with transporting equipment and inventory to a safe location.

Also as of late Monday evening—

—No injuries from the flooding were reported.

—Flooding was mainly limited to the lower end of Richwood and to the Riverside Addition areas.

—The city's public water supply was not contaminated and remained safe for consumption.

Sutton expressed thanks for the new storm sewer which he credited with saving the downtown area from damages. As a first real test, it has done more than many had anticipated would be necessary for the system.

Rural areas hit by flood

Nov. 7
Mountain Statesman
Grafton
By Bruce Carter

"It wiped us out, there's nothing left," is how one Thornton area resident summarized the effect of flood waters on his family earlier this week.

"We were sitting there watching television about nine Monday night and I looked outside. The water was all around the house. By the time we could get everybody together, it was coming into the house. There was just no warning that it was rising that fast.

"We went over to a friend's house that is up across the road. When we came back Tuesday morning there was three feet of water in the house," he said.

MARLINTON—Road damage at Minnehaha Springs. (Courtesy of Pocahontas Times)

Flooding: recovery begins

Nov. 13
Tyler Star News
Sistersville

Relief in the form of donations of hundreds of dollars, tons of goods and thousands of hours of labor was getting to Tyler County flood victims this week.

Tyler County had been included in the section of West Virginia designated by federal officials as a "disaster area." Though it was expected that designation could provide federal aid to local flood victims, few plans to provide that help had been made by Tuesday.

Contributions of money, non-perishable foods, appliances, furniture, cleaning supplies and bedding were being sought this week by a group of organizations and individuals. Donations were being used to aid Tyler Countians who suffered losses during last weeks's flooding.

A spokesman for the relief organizations said more than enough donations of clothing have been received. Needed now are the items listed above, he stressed.

Once the needs of local flood victims are met, any donations remaining will be used to aid those in eastern West Virginia, even harder hit than the local area.

Tyler County Sheriff Gary Keller on Tuesday estimated that 80-100 families suffered from flash flooding in the local area last week.

Keller on Nov. 5 had been concerned about two pregnant women in the Shirley area. The women, due to give birth "any time now," were cut off from medical help by flooded roads.

He said he discussed the situation with the Governor's Office of Emergency Services. "They said, 'If you want a helicopter (to take the women to medical care facilities), just say so and you'll have it'," Keller reported. He said the Army National Guard told him "a helicopter was only half an hour away if I needed it" for the women.

On Nov. 6 Keller noted roads to Shirley were open.

Of the two women, he reported on Tuesday that, "As far as I know, they're still pregnant."

Keller said there were no reports of serious injuries caused by Tyler County flooding, though "we had several close calls. I know of at least five vehicles with people in them and water got them and they had to leave the vehicles to get out alive."

"A number of homes were heavily damaged. The water got in there and buckled the wood. It's in the foundation; that's major damage," Keller explained.

Dozens of families in Tyler County lost personal belongings because flood water rose so quickly there was no time to remove furniture, appliances, or even clothing.

One disaster relief agency planned to establish a center in Clarksburg, to which people in need of help could go to request it. Keller said that center was to serve several counties, including Tyler.

"They'd just as well set it up in Michigan," Keller said of the center, explaining that "We've got a lot of poor people and a lot of old people who can't travel" that distance to get help.

The Middlebourne Volunteer Fire Department was being used as

MARLINTON—Damage was extensive to the interior of the Marlinton Elementary School. **(Courtesy of The Pocahontas Times)**

a center for collection and distribution of goods donated to help local flood victims. Other sites throughout the county were collecting items and sending them to Middlebourne.

Middlebourne Fire Chief Mitch Wilcox said volunteer members of his department had, by Tuesday morning, spent a total of 1,948 hours helping flood victims. Members of the MVFD's junior fire department—some as young as nine years old—had worked an additional 163 hours and members of the department's auxiliary were helping, Wilcox said.

"Some of these guys went three days without sleep," Wilcox said of his department's work.

He added that the fire department "would like to thank all the people who helped us out" with labor and donations.

Areas hit hardest by flooding in Tyler County were in the eastern and central sections of the county.

Floods on Nov. 4-6 closed roads, isolated several small communities, caused some power and telephone outages and shut down schools. Tyler County schools, dismissed early on Nov. 4, were closed Nov. 5-6.

Several farmers lost large amounts of hay and other crops and a few had expensive farm equipment damaged.

Keller reported at least six vehicles sustained extensive damage, while several others received minor water damage. Included in the list of water-damaged vehicles were four or five surplus school buses which had been sold by the county school system but were parked at Tyler County's school bus garage on Nov. 5. New owners of the buses were not able to move them in time to avoid flooding.

About four camping trailers and a gasoline tanker trailer were washed into creeks.

Flood water hit at least one Monogahela Power Co. substation and several other facilities. Power lines and utility poles in several areas were downed by the flooding.

Chesapeake and Potomac Telephone Co. facilities sustained similar damage.

Crews from both utility companies were kept busy both last week repairing damage and restoring service to customers.

Keller said some oil drilling rigs and wells also were damaged.

Between 30-40 homes in Middlebourne were flooded and Keller said he counted 51 flooded residences along W.Va. Route 18 between Middlebourne and the Tyler-Doddridge County line.

Homes in several other areas of the county also were flooded.

Though the effects of flooding in Tyler County were not as drastic as in some other sections of the state, Keller said many families suffered severe financial losses.

"I know all these people don't have flood insurance," he said, adding "Some of them couldn't afford it anyway. A lot of people aren't working; they're just devastated."

The sheriff said state officials had been extremely helpful in offering aid to Tyler County.

MARLINTON—The media was not spared the wrath of the flood. The front room of The Pocahontas Times offices was the scene of considerable damage. (Courtesy of The Pocahontas Times)

SENECA ROCKS—The North Branch of the South Fork of the Potomac River has entirely changed its course in Pendleton County. (Courtesy of Pendleton Times)

FRANKLIN—Trailers at the Cave Country Campground were pushed aside by the waters of the North Fork. (Courtesy of Pendleton Times)

"The town of Hendricks is done for."

—— Mabie resident Faye Bertelli on seeing the town

We have a job to do

Nov. 18
The Pendleton Times
Franklin

Pendleton County residents are experiencing a crisis unequaled in this county since the Civil War. Lives have been lost, homes have been destroyed and it appears that the very economic foundation of the county has been threatened.

Many are bewildered, hungry and cold. Many have lost everything they owned—their homes, their automobiles and all their possessions. For the first time in their lives many of our citizens are destitute.

But Pendletonians have suffered before. Many of our citizens fought and suffered in World War I and World War II, in the Korean Conflict and in Vietnam. Many of our older residents remember the fire that destroyed Franklin in 1924, and many more remember the floods that wrecked havoc in 1936 and 1949.

The people of Pendleton County are a rugged bunch. They are not strangers to hardship or frustration or hard work. They perhaps are more familiar with the ravages of life than with life's luxuries. And their background and heredity will stand them in good stead in the present crisis.

Through hard work, perseverance and dedication, our ancestors forged a home here in the woods and hills and valleys of Pendleton County, and our parents made it better. In spite of what the flood has done to our beautiful county, we still are better off than were those before us. And with the personal qualities they have passed on to us—the ability to think and to dream and to work—we will repair the damage.

We have witnessed in our local people during this crisis all the qualities that are required to overcome our present hardship. The ability of our people to cope with difficult situations, and the willingness to cooperate and help one another has been a reassuring sign that we will not give in to this disaster.

This is no time for despair and hopelessness. We cannot afford to feel sorry for ourselves. We must stick together and do what is necessary to clean up the debris, to return the streams and rivers to their channels and rebuild our homes and farms. Above all, we must not become discouraged. We must not think the job is hopeless.

With the personal resources that God has given us, we will rebuild our county and make it better than ever. We can do it. We will do it.

30 miles of farm land will never be farmed again in Pendleton County

Nov. 8
Inter-Mountain
Elkins
By Tom Cater

"30 miles of land as wide as the valley will never be farmed again," said Agricultural Commissioner Gus Douglass upon returning from an aerial damage assessment of farms in the flood torn communities of Pendleton County.

"It's unbelieveable," Douglass said. "The land looks like the bottom of a river. All the top soil has been washed away and there is nothing but rocks and debris scattered everywhere, roofing, timber, household items, it is unbelievable."

Douglass said that he had attended a meeting earlier in the afternoon and was told that more than 2,000 cattle, 1,000 hogs and hundreds of thousands of turkeys had been killed in the flood. "We're trying to get equipment in there to bury the dead animals, but there are no roads," he said.

"Tractors were buried in the mud, with only the tops sticking out. We must have seen a million dollars worth of farm equipment buried in the mud."

FRANKLIN—Floodwaters buckled the Mill Creek Road. (Courtesy of Piedmont Herald)

UPPER TRACT—Marked for demolition this flood damaged Pendleton County business sits waiting for the bulldozers. (Courtesy of Piedmont Herald)

FRANKLIN—This Franklin home ended its watery trip in the middle of Mill Road. (Courtesy of Piedmont Herald)

RIVERTON—The community building in Riverton found one wall caved in. (Courtesy of Pendleton Times)

ONEGO—Trailers on Brushy Run were destroyed by the power of the North Fork. (Courtesy of Pendleton Times)

FRANKLIN Debris left by the flooded South Branch of the Potomac River was strewn throughout Franklin. (Courtesy of the Inter-Mountain)

CIRCLEVILLE—This barn was toppled by floodwaters from the North Fork. (Courtesy of the Inter-Mountain)

MILLVILLE—Linda and Calvin Mitchell, who lost their mobile home in flooding on the Shenandoah River at the Bloomery Bridge on W.Va. 9, make a last ditch effort to save some items. The waters there crested at 26 feet about 9 p.m. Nov. 6. (Courtesy of Martinsburg Evening Journal)

Damage estimate doubles

Nov. 9
Evening Journal
Martinsburg
By Todd Spangler

Harpers Ferry Park officials estimate damages at close to $500,000 since the water level dropped below flood stage at 6 a.m. Friday, but part of the historic district may still be open by early next week.

"Where we were predicting approximately $250,000 damage, now we're looking at closer to $500,000," said Donald Campell, superintendent for the National Park Service at Harper's Ferry.

Campbell said additional damage has been discovered since the water level dropped to a stage where workers could enter the affected buildings on Shenandoah Street and assess the destruction.

"The biggest surprise was the parking lot in the lower town," Campbell said. "The runoff current created a heavy channel that cut through the middle third of the parking lot. It looks like pieces of peanut brittle."

The Park Service will be working throughout the weekend, attempting to clear mud out of buildings and restore them to workable order. None of the relics, moved to higher ground before the flood reached Harpers Ferry, have been returned to any of the buildings.

Campbell said a temporary visitor center will be established on High Street in an unaffected building above the Pharmacy. Few exhibits are expected to be open if the park is open in a limited capacity next week.

Walls, street lighting and outdoor landscaping will also have to be repaired.

"It did a lot of damage on Shenandoah Street," Campbell said. "The current was 20 knots or more."

A portion of Harper Ferry's sewer system was also shut down by the flood. The pump station, located in the parking lot in the historic district, was switched from an electrical system to a diesel engine when the flood covered the switch box.

The diesel engine, according to Dixie Kilham, chairman of the Harpers Ferry Public Service District, was flooded and service to the lower third of the town was stopped.

Kilham said he does not know how much the damage to the pump station will cost.

"It's not a little problem," he said, "but it's no cause for hysteria."

Kilham said sewer service should be restored to the affected part of Harpers Ferry today.

HARPERS FERRY—National Park Service buildings in the historic area were flooded to the first floor by the raging Shenandoah River early on Nov. 6. The river crest of 29 feet was expected, reaching the level of the flood caused by Hurricane Agnes in 1972.

Jefferson spared serious damage, relief is offered

Nov. 14
Spirit of Jefferson Advocate
Charles Town

One of the three worst floods in the last half century left devastation and despair the length and breadth of the Potomac and Shenanhoah River watersheds last week.

Headwaters of the two streams, which merge at Harpers Ferry, were deluged by torrential rains that brought eight or more inches in less than a twenty-four hour period.

The downpour created flood crests which swept through many communities, and the surge carried all the way to Harpers Ferry and beyond before abating and sparing, to a large extent, the D.C. area.

Observers point out that the flooding would have been worse had there been heavy rainfall in the downstream areas of both the Shenandoah and Potomac rivers. Here in Jefferson County, weather observer Laurence Lloyd, Sr., measured the rainfall at less than three inches over a four-day period. And small streams, while bankfull, never really left their banks.

Always Trouble

The Bloomery, always a trouble spot, was hardest hit. There, at three trailer parks, Wares, Prices and Jones, some twenty-seven mobile homes were either destroyed or badly damaged. Across the river, at Hobo Flats, not yet fully surveyed, seasonal campers, mobile homes and cabins were hard hit by the cresting Shenandoah.

At Millville, seasonal cabins were flood victims as was Whipp's Grocery and the Pentecostal Church. Railroad tracks to the Millville quarry were also washed out.

Penwell reported that actual damage on the Potomac was well below that on the Shenandoah. At Shepherdstown, there was minor damage along the River Road. Glen Haven sub-division sustained very little damage; and at Harpers Ferry damage was confined to National Park Service structures on Shenandoah Street and to the parking area of the park. The dollar loss was, however, the highest of any of the areas, with park officials now saying clean-up and repair will be approximately a half-million dollars.

Flood warnings were issued Tuesday, November 5, and evacuation of low-lying river areas was begun.

Penwell also reported the almost complete absence of injury. The only report was that of a young *Baltimore Sun* girl reporter who, after spending hours on the flood scene at Harpers Ferry, fell asleep. When she awoke and rubbed her eyes, one eye was scratched by a contact lens.

The flood crested on both rivers in Jefferson County late Wednesdy, Nov. 6, 31.4 feet at Shepherdstown, more than 16 feet above flood stage; 33.5 at Harpers Ferry, 15.5 feet above flood stage; and 25.7 feet at Millville, 12.7 feet above flood stage.

Map strategy for searching cars; removing animal carcasses

Nov. 8
Inter-Mountain
Elkins

A strategy meeting on the massive task of cleaning up thousands of dead animals, and searching cars swept away by flood waters for survivors was discussed at a meeting of the Elkins-Randolph County Emergency Services headquarters today.

Coordinator Ashby Craft said now that the flood waters have receded, there are still a number of missing, and services must be coordinated to provide for the needs of those people who have lost their homes in Randolph and neighboring counties.

One of the largest health problems facing emergency services efforts is the staggering amount of work that will be required to remove the thousands of dead animal carcasses strung along streambanks for miles in Pendleton, Tucker and Randolph Counties.

Over 2,000 cattle were drowned in the flood, and many of them weigh close to two thousand pounds. "It will be difficult to move them, to dig them out of the mud and bury that many animals before they begin to putrify and become a health hazard," said an emergency services worker.

Another problem facing salvage projects are the hundreds of thousands of turkeys that need to be picked up and buried.

HARPERS FERRY—Workers scramble to clear an aquaduct of debris to facilitate drainage away from national treasures.

(Courtesy of The Evening Journal)

BLOOMERY—Shenandoah River water continued to rise into homes and roadways, forcing many residents to flee. (Courtesy of Jefferson Advocate)

JEFFERSON COUNTY—The Shenandoah River spilled over its banks and inundated homes near the Bloomery Bridge. (Courtesy of Evening Journal)

ALBRIGHT—The tragedy of this Preston County town is revealed in a sweeping view of the devastation. (Courtesy of Exponent Telegram)

Flood destroys at least 193 homes, Preston declared national disaster area

Nov. 12
Preston County News
Terra Alta
By Greg Hunter

The recent flood is now just a vivid nightmare in the minds of thousands of Preston County residents who live along the Cheat River. All that is left to do now is to pick up the pieces and count the losses.

Those losses could be staggering. Seventy-seven homes were reportedly totally destroyed in the Rowlesburg area, 38 more severely damaged. Sixty homes lost in Albright, 15 heavily damaged. Fifty-six homes are gone in the Macomber area, 17 were more badly damaged. Along with those losses are three bridges, countless miles of roads and railroad track, one electric power plant substation in Albright, numerous businesses, and, tragically, at least one life.

Lynn Taylor of the Preston Country Club was found dead last week. His body had been swept all the way to Cheat Lake in Monongalia County. Also, Josephine Newcome is missing and feared dead as her trailer was washed away in Albright, though no body has yet been found.

In the state, 26 people have now been listed as dead by one of the worst disasters in West Virginia's history. Another 29 persons are said to be missing and hopes of finding any of those alive decreases with each passing day.

Twenty-nine West Virginia counties have now been declared federal disaster areas by President Ronald Reagan. Preston was one of the original eight the president declared last Thursday. The declaration clears the way for the Federal Emergency Management Agency (FEMA) to start the process of providing federal aid in temporary housing, clean-up operations and a number of programs available to flood victims.

Hundreds of jobs have also been affected by the flood even though the job sites may not have been directly hit by the water. The coal mining industry in Preston County has been almost paralyzed because of the loss of the railroad trestle at Rowlesburg, according to Bob Allen, director of Tri-State Coal Operators Association.

"Our main goal is to try to keep everyone working," said Allen. "This county doesn't need layoffs on top of all the other problems we have had. There will probably have to be a few layoffs for a while, but Tri-State has worked out a schedule for all the coal operators to ship their coal out of the county. It may be an inconvenience to the public at times, because we will have to block some roads to fill the railroad cars, but I hope people remember that it is what is keeping 947 coal people working in this county, not to mention all the truck drivers and everyone who depends on the industry for their livings also.

"Hopefully we won't have to use our alternate schedule for too long. Chessie Railroad has told us they expect to have the Rowlesburg railroad trestle back in operation by Nov. 18."

WVU has sent up 130 of its physical plant workers to assist the Rowlesburg, Albright and Kingwood water and sewage workers to get those plants back on line.

Rowlesburg residents trying to put the pieces of their lives together

Nov. 14
The Preston County Journal
Kingwood
By Linda Benson

"Hi, neighbor," Pat Likens called to Vera Hayden Saturday. "I hope we'll be neighbors again."

The two women had been neighbors until the Nov. 4 flood roared through Rowlesburg. Now all that remains is a hole where the Likens house used to stand. The Hayden house was not completely demolished, but it was torn from its foundation, its windows were broken and mud "two feet thick upstairs" hampered efforts to recover any belongings that remained inside.

"The fireman told us (she and her husband John) the water was coming up," Mrs. Likens said. And when the B and O bridge collapsed it diverted water through her section of town making a "clean sweep". Where houses stood all that remains are holes or cement slabs that used to be the foundation of homes.

Mrs. Likens believes she and her husband will rebuild on the property.

"That is all we had. We worked all those years for this," she said.

Mrs. Hayden, who is staying with a friend, is not sure about her future plans.

"It's too soon," she said.

Other neighbors were out working in Rowlesburg Saturday, too. One woman tried to follow the trail her home took when it vanished in the flood. She found a piece of material for a skirt she had been working on before the flood. The material was tangled in a tree. She and her daughter walked on looking for more clues that would tell her what had happened to her home.

Alice Braithwood returned to what was left of her home on Saturday.

She recalled the night of the flood.

The water began to rise around 8 p.m., she said. At 9:30 p.m. her neighbor Mary Short called and told her she had gotten a call from a local volunteer fireman. He had told her to go out and tell the neighbors about the rising river, Mrs. Braithwood said.

Mrs. Braithwood grabbed some papers and her pocketbook, got

ROWLESBURG—A Rowlesburg family views the devastation where their home once stood. (Courtesy of Fairmont Times-West Virginian)

into her car and went up on Main Street. At around 1 a.m. she headed for higher ground.

She has now rented an apartment in Kingwood. She is bitter that she had no more warning about the flood.

"I've lived here 67 years," she said. "We used to be notified. This time we didn't know until it was almost too late."

She said she will never again live on low ground near the river.

"I care about my house and 45 years of things, but I lost one of my cats," she said as tears filled her eyes.

A lot of people came back to Rowlesburg after the flood to help their relatives.

On Saturday John Marshal of Hampton, Va., was busy shoveling mud from the basement of the VFW. His wife was one of the Rowlesburg Farises and they came back home to help.

Mrs. Marshal's mother was one of the lucky ones, the water didn't get up into her house.

George Lipscomb lives up away from the flood, but he was in town Saturday, also working to help clean the VFW.

Jerry Felton, who has relatives in Rowlesburg, came from Short Gap to help with the cleanup. On Saturday he was busy working at Frank Funk's house. The house was destroyed, but a few things were saved. There was a cedar chest filled with quilts, and most of the quilts were safe. Mr. Funk also found a railroad watch with a fob that was the last gift he received from his late son.

A few other things that were saved were stolen by looters after the flood. He and others who lost some of the little they were able to retrieve from the floodwater are angry that such scavengers could further prey upon the town.

On Saturday Frank Funk looked at the remains of his home.

"I'm back to where I was 66 years ago," he said.

He recalled some of the things he had lost, antique dishes; a clock given to him by Dr. Proudfoot for which he had been offered $800.

He and his wife were among those who had the most difficulty getting out of the flood's path.

Mr. Funk's niece, Wanda Wiles, warned them about the approaching flood. Mrs. Funk is very ill and has to use oxygen and other life support systems. Mrs. Wiles took Mrs. Funk to her house, but when they reached there the electricity went off. Electricity was needed to keep the support systems operating.

Mrs. Frank Funk was then taken to the home of Mary McMillen,

ROWLESBURG—An aerial view of the path the Cheat River scoured out in Rowlesburg. (Courtesy of Clarksburg Exponent Telegram)

ROWLESBURG—Residents along the Cheat River cleaned out the muddy, destroyed items from their homes. (Courtesy of Dominion Post)

but just as she arrived the electricity went off there too. The Funks stayed at the McMillen home all night, however, and the life support was kept operating with an emergency generator. The next morning a national guard helicopter airlifted the Funks to Kingwood.

Mrs. Wiles' husband John and daughter Debbie Beveridge in the meantime came back down into Rowlesburg to get a portable oxygen tank and a TV from the upstairs of the Funk home. When they got back to the street the water was thigh high.

Around 2:30 Debbie went to the house of Phoebe Wiles, 91, on the South Side of town. She saw the water coming up Walnut Street and had to run the whole length of the porch, she recalls. She woke Mrs. Wiles and two other families in the apartment building. All had gone to bed at 10 p.m., not knowing anything was wrong.

"We were running and the water kept coming," Debbie recalls.

But all escaped, a fact Debbie finds hard to believe.

The people of Rowlesburg recall the night of the flood, how people kept retreating as the water kept coming.

"You couldn't get another person on the hill. People stayed in their cars," Mrs. Wiles remembered.

But the nightmare night is over and now the families must get on with their lives.

Some plan to stay in Rowlesburg, some plan to go other places, as least temporarily. And some don't. As Mrs. Hayden said. "It's too soon."

No cake for Nathan

Nov. 7
Preston County News
Terra Alta
By Bob Teets

Albright resident Marion Custer, whose home was unaffected by the flood waters, spent most of the day Tuesday at the school helping out.

She did report one loss, however.

Her son Nathan had turned six that morning.

She had had a friend prepare a birthday cake for him.

"When he woke up, he wanted to know where his birthday cake was," she said.

It had gone downriver with the friend's home.

"But when this is all over," she told Nathan, "you're going to get a super cake."

Army copter rescues father, son from river

Nov. 6
Dominion Post
Morgantown
By Nancy Abrams

A Preston County father and son were plucked from a tree in the flooded Cheat River early Tuesday morning by an Army helicopter.

Fred Snider, age unknown, and 6-year-old Adam were in a barn at their campground on an island near Route 50 at Rowlesburg when the river began rising Monday night.

Norma Snider of Kingwood, Fred's mother, said the family was in phone contact with Fred and Adam until about 2 a.m. Tuesday. The two had taken refuge in the loft of the barn.

"That was the last time we were able to talk to him," Norma said, hugging Adam on her lap in the laundry room of Preston Memorial Hospital. "Fred said then he was going to build a raft and get to a tree."

Adam told how they built the raft. "We had two great big coolers and we tied them together with wire and floated about half

ALBRIGHT—The raging Cheat River tears into the Rt. 26 bridge at Albright. (Courtesy of Preston County News)

ALBRIGHT—Entire trees were tossed about like twigs. This one came to rest on top of the Rt. 26 bridge which spans the Cheat River. (Courtesy of Dominion Post)

a mile." Clad in mismatched slippers, the boy spoke for his father, who declined to talk.

"We were in the tree for seven hours," Adam said, acknowledging he had never been as cold as he was Monday night.

An Army helicopter spotted the two around 8 a.m. in a tree and first lowered a rope to move the boy to dry ground. The father was rescued next. After reaching ground, the two were loaded into the helicopter and flown to Preston Memorial Hospital.

Fred's wife, Barbara, and daughter, Sarah were reported safe in Rowlesburg. Norma said Barbara probably learned of the dramatic rescue on the radio, as phones were not working in the flooded riverfront community.

ROWLESBURG—The X means that this flood damaged Manheim home has been condemned. (Courtesy of Dominion Post)

ROWLESBURG—The twisted remains of the railroad trestle that once spanned the Cheat River. (Courtesy of Dominion Post)

Flood didn't take everything

As it appeared in *The Wheeling Intelligencer*

By Adam Kelly

"They lost everything."

That grim assessment is heard often in West Virginia these days. This family; that business, "lost everything." The statement has been heard on the street below the high school in Rowlesburg and on Main Street in Albright and Petersburg and in Franklin and Belington. It has been heard, in fact, over almost half of the geographical area of our state. "They lost everything."

The comment was made on Pennsylvania Avenue.

Pennsylvania was a quiet, tree-lined residential thoroughfare in a quiet, residential Tucker County town called Parsons. But the devastating flood removed all the trees and most of the houses from Pennsylvania Avenue. There isn't much left.

"They must have lost everything," said people standing in the wreckage of one home and looking across Pennsylvania Avenue to what was left of another.

One entire end of the building which had drawn their attention had been sliced away cleanly, as if by some giant wielding a can opener. All the rooms ruthlessly had been exposed to view. Broken pieces of furniture were scattered about. It was as if that same giant had tossed down the pieces in a sudden fit of anger, like a child with her doll house and play furniture.

"They lost everything."

No.

For, stretching almost clear across where one of the rooms upstairs once had been, its rippling folds mercifully obscuring a sad vista of personal tragedy, was an American flag. It was a huge flag. The people who once lived in that house had taken the time, in the midst of their own great personal disaster, to fly the flag of their country for Veterans Day. Lost everything? Not the people who live on Pennsylvania Avenue in Parsons, W.Va. For they have an abiding faith in their country.

Water reached the brow of the figure of Our Lady of Mercy at the front of the Catholic Church in Parsons. It almost seemed as she were weeping bitter tears at the desolate scene unfolding before her in what once had been the business district of the town.

But high up on the hill, where the waters couldn't reach, there

ALBRIGHT—Albright flood victims search through their battered homes in an attempt to find anything salvagable. (Courtesy of Dominion Post)

was laughter and warmth, and light, and the smell of good food and the offer of shelter for the night. The Catholics and the Baptists and the Presbyterians and folks from all the churches—and some good people who own no denomination—all were using the undamaged United Methodist Church to do the work of the Lord.

Inside, maybe some of the busy workers remembered, as they cooked so many meals, how He once said, "I was hungry and ye fed me." And maybe they didn't because there wasn't much time for reflection, what with so much work to be done and so many hungry people to be fed.

Anyway He would understand because He certainly knew what it meant to feed a multitude. And One who was born in a stable and later said, "I was a stranger and ye took me in," would know how important it was to get shelter over the heads of the suddenly homeless.

So a family of four, parents and two children, came in for food and shelter. They found it there, and love as well. The dazed expression in their eyes revealed the tragic story of how hard it was for them even to begin to comprehend that their house and all that was in it was gone. Gone. ("They lost everything," someone whispered.)

The family members bowed their heads, in grateful acknowledgement of their food. A sudden hush came across the room and the little girl's voice was heard plainly: "And thank you, God, for everything." Some whispered, "Amen." Others wiped away tears. And the realization came. These did not lose everything. They never can lose all for such as these have an abiding faith in God. And with that, all never can and never will be lost.

People such as these know that He is there with them in Parsons and Franklin and everywhere. They know that even in tough times such as these His love for them is present. It's being expressed through the love of people for people. Anyone who doubts the expression of that love can just count the number of trucks loaded with food and clothes and furniture rolling through the mountains.

Faith in God, faith in country, faith in the love of people for each other. With these, none ever "loses everything." Not in Parsons or Franklin or Rowlesburg, or anywhere.

Everybody has a flood story

Nov. 11
The *Record Delta*
Buckhannon
By Barbara Clowdus

The day of the flood is stamped forever on the consciousness of West Virginians as indelibly as the day John F. Kennedy was shot.

People will remember always where they were and what they were doing.

These are the thousands of individual stories that hum through

ALBRIGHT—The force of the Cheat River impaled this chair into this home. (Courtesy of Dominion Post)

PHILIPPI—Main Street at approximately 1 p.m. on Tuesday, November 5, as flood waters began receding. Philippi Volunteer Fire Department boat is pictured as residents are removed from apartments in the Alkire building. (Courtesy of The Democrat)

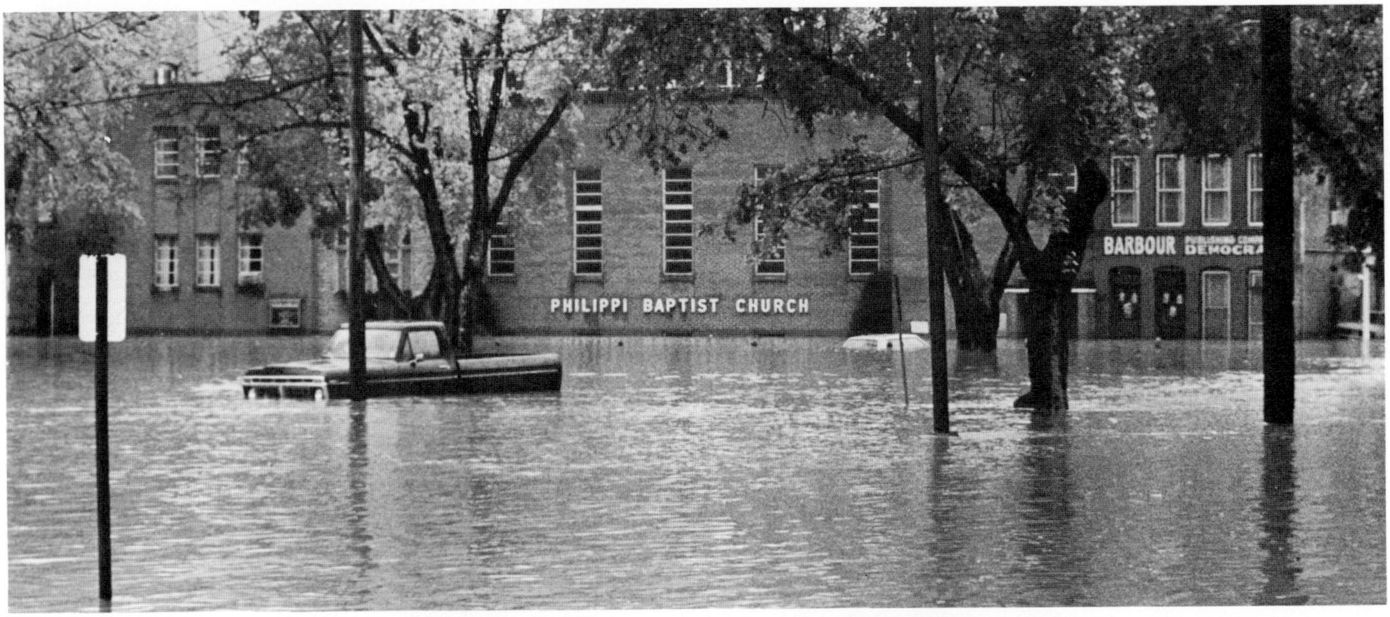

PHILIPPI—Floodwaters poured through all parts of town, inundating churches and businesses. (Courtesy of The Democrat)

the county as stricken families salvage what they can of businesses and homes and personal momentoes that seem to verify life.

Upshur County was spared the total devastation of neighboring counties. Here, it is primarily a gigantic cleanup job.

Junior Heater of French Creek tried to retrieve his office trailer from the now-tame river that flows beside his trucking business on the Gould Road in Adrian. Resisting the attempt, the trailer collapsed in the river bed, one-quarter of a mile downstream.

"I got all my trucks outa here," Heater said, "but who would've thought the water would've gotten high enough to carry that trailer off like that. I would never have believed it."

That sense of increduality, of being caught by surprise, is a common thread that has woven a costly blanket.

And as the death count mounted in the state and in Virginia, the thump, thump of the National Guard helicopters overhead ferrying medical supplies and water seemed like heartbeats uniting Upshur County with flood victims everywhere.

That awareness has gendered a feeling of thankfulness, of resolve, of determination, and optimism here, which has surfaced along with the mud.

"Oh well," said many people. "It could have been so much worse here."

And Junior Heater contemplated for a short time, he said, the possibility of moving his business away from the river.

"Oh well," he said, "I might as well stay here. I figure since this is a 100-year flood, I won't be around for the next one anyway."

Terri Chanell, of the circuit court clerk's office, said she waded through water nearly to her waist on Reger Street Monday night to get home to her children. Her husband remained stranded in Clarksburg for two days.

Other mothers, like Ellen Lake of Central National Bank, remained stranded in town, with her children 12 miles away at home.

Business owner Neale Thacker of Triangle Service Station, used to contending with high water, also was caught by surprise.

Thacker said he got many of the cars, but not all, at this service station moved to higher ground.

PHILIPPI—It has withstood the tromping of Civil War soldiers and the thumping of modern day traffic. On Nov. 4-5, the Covered Bridge took a battering from the river it was built to conquer. And although it lost some facing, it remained standing when the waters of the Tygart fell away. The symbol of West Virginia's heritage has now also become a symbol of Mountaineer strength. (Courtesy The Democrat)

BELINGTON—Rain-swollen rivers throughout West Virginia affected neighbors, and neighborhoods, alike, as shown in this section of town. (Courtesy of Inter-Mountain)

BARBOUR COUNTY—It was reported that 99 percent of the county's businesses suffered extensive losses because, like residents, merchants did not have time to move belongings before the waters poured in. (Courtesy of The Democrat)

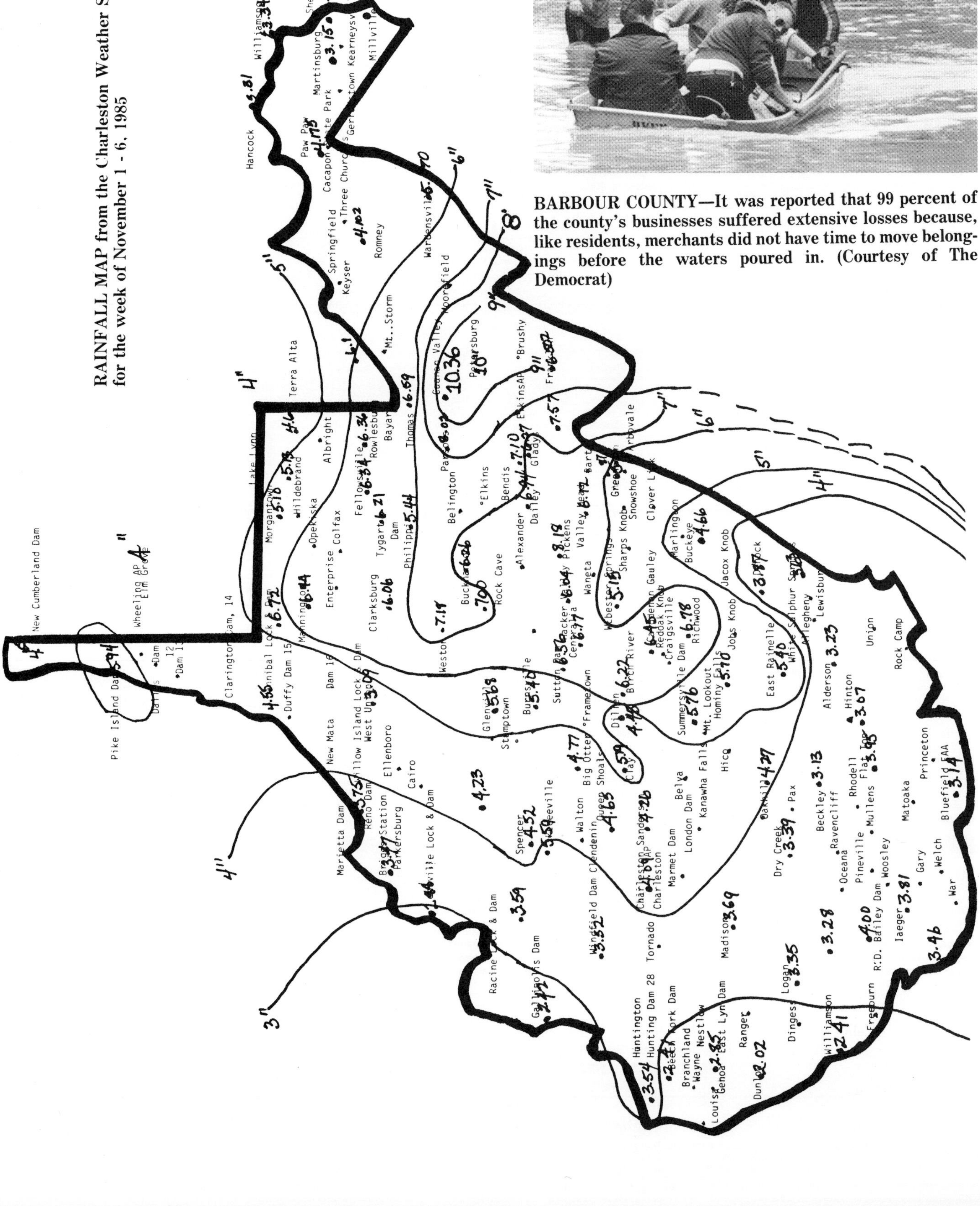

RAINFALL MAP from the Charleston Weather Service for the week of November 1 - 6, 1985

PHILIPPI—Joe Pitman is evacuated from his home and helped into a row boat by Curtis Liever, left and Rodney Sturm Nov. 5 on Main Street as flood waters crested. (Associated Press Wire Photo Courtesy of Huntington Herald Dispatch)

Most county businesses and one third of residences devastated by flooding last week

Nov. 13
The Barbour Democrat
Philippi

Literally 99 percent of the businesses and at least one-third of the homes which house the population of Barbour County were affected, if not totally wiped out, in the worst natural disaster ever to hit West Virginia, when flood waters raised to an all-time height, early Tuesday morning, November 5, at approximately 3:30 a.m.

Hundreds of residents of Philippi, Belington and Junior decided to leave their homes earlier Monday evening, others were forced to evacuate in the early morning hours on Tuesday. Many waded flood waters waist-deep, to waiting boats, salvaging only the clothes on their backs. Both Barbour County and Belington emergency squads and Philippi, Junior, and Belington volunteer firemen worked around the clock, giving emergency assistance to residents.

The Tygart Valley River crested around noon on Tuesday, forcing flood waters into the streets of Junior, Belington, and Philippi, covering bridges and roadways into and out of the county, leaving some people stranded in their homes, while others were unable to get home and were separated from their families. Cut off from gas, electric, water, and phone service, people were taken to emergency shelters being set up at various points throughout the county. Emergency squads and fire departments were still responding to calls to help some 48 hours after the peak of the disaster.

In two major floods in 1912 and 1967, flood waters did extensive damage to county property when the Tygart Valley River crested at 26 1/2 feet. The weather bureau at Elkins and Washington, D.C., reported the official word on the 1985 flood, last week, as flood water crested at 31 feet, nine inches. Automatic readings were checked regularly until waters raised to the 26-foot mark. At that time the automatic mechanism in Philippi was washed away by the swift flood waters. At that point Philippi city police started taking readings every hour on the hour.

Waters raised briskly, at an unbelievable rate, leaving residents no time for preparation.

Philippi was virtually cut-off from U.S. Routes 119 and 250, both on the northern and southern ends of the city. Waters receded enough by Thursday morning to render the new bridge in North Philippi passable around 7 o'clock. The old historical Philippi Covered Bridge was not made passable until Friday, November 8, at 5 p.m.

National Guardsmen were in the Philippi area as early as 7 a.m. on Tuesday. Other troops and equipment were brought in Wednesday to assist in controlling the traffic and to patrol the area of any unnecessary looting. National guard units and equipment aided in the clean-up, which began Friday morning, when residents and merchants tried to recover from what looked like a violated war zone.

National Weather Service Flood Warnings

	TIME ISSUED	DATE	LOCATION	COMMENTS
1	4:30 am	Mon. Nov.4	Charleston	Flood Watch - Pendel. Grant, Hardy, Mineral, Hamp,MOrgan, Berkley, Jefferson Weather service did not expect a rapid rise in water levels as yet
2	10:50 am	"	Elkins	Flood Warning - expires 4 pm Nov. 4. Small stream flood warning in Randolph
3	10:30 am	"	Charleston	Flood Warning - exp. 3:00 pm - lewis, gilmer, Braxton
3	11:00 am	"	"	Flood Warning - exp. 4:00 pm - Doddridge, Ritchie, Tyler
4	12:00 noon	"	"	Flood Statement issued -Flood watch in Potomas Highland & Eastern Panhandle Small stream warning for Tyler, Ritchie, Doddridge, Lewis, Gilmer, Braxton
5	?	"	"	Flooding Forecast along the West Fork River
6	1:00 pm	"	"	Flood Warning - Wetzel, Pleasant, Clay, Calhoun, Nicholas, Webster, Greenbrier Stream overflow already in Greenbrier, Nicholas, Gilmer, Clay, Lewis, Dodd.
7	1:15 pm	"	Elkins	Flash Flood Warning exp. 5 pm. affecting Randolph & Pocahontas
	?	"	Charleston	Flood Warning Cherry, Cranberry, & Williams Rivers (Nicholas & Webster
	1:30 pm	"	Elkins	Flood Warning exp. 5 pm. Hardy & Upshur
	2:00 pm	"	"	Flood Warning exp. 5 pm. - Grant - Water starting to cover highways
9	2:00 pm	"	Charleston	Water & Rainfall Statement * highwater on the Little Kanawha River*
10	3:30 pm	"	"	River Flood Warning - Greenbrier River
12	4:45 pm	"	Charleston	River Flood Warning * CHEAT river to include the Shavers Fork & Blackfork
14	6:00 pm	"	"	River Flood Warning - MAJOR FLOODING forecast for the Tygart Valley River
13	6:00 pm	"	"	River Flood Warning for the Little Kanawha River
11	6:30 pm	"	"	River Flood Warning No. 2 * Record Flooding forecast along the West Fork River (Weston/Clarksburg)
15	7:30 pm	"	"	River Flood Warning 2 - Revised crest forecast for Little Kanawha River
17	8:30 pm	"	"	River Flood Warning 3 - Revised Crest Forcast for Little Kanawha River
16	?	"	"	River Flood Warning - Flooding forecast along the Gauley River (Summersville)
20	8:59 pm	"	Cincinnati RFC	River Forecasts (flood stage & crests)
18	9:00 pm	"	Charleston	River Flood Warning Number 2 - Tygart River flood warning update
19	9:30 pm	"	"	Cheat River Flood Warning Update - Parsons/Rowlesburg
21	10:11 pm	"	Cincinnati RFC	River Forecasts - Alderson
20	11:15 pm	"	"	River Flood Warning 4 - Record flooding forecast for Little Kanawha River
21	12:15 am	Nov. 5	Charleston	River Flood Warning Number 3 - Record flood on the Tygart River (Belington/Phil
23	1:00 am	"	"	River Flood Warning Number 2 - Record flooding on the Greenbrier River River flood forecasts for the West Fork, Cheat, Little Kanawha, Tygart will be listed in separate statements
25	2:00 am	"	"	Flood Statement - Evacuation shelters set up in North Central & Eastern area Hardest hit - Weston, Marlington, Philippi, Burlington. Record flooding forecast along the West Fork, Greenbrier, Cheat, Little Kan. & Tygart
25	2:00 am	"	"	River & Rainfall Statement
27	?	"	"	Flooding cont. along the West Fork River
27	5:45 am	"	"	River Flood Statement - Tygart River Crested at Dailey, WV, Cheat River crested at Parsons, WV about 3 am at an unknown stage
28	6:45 am	"	"	River FLood Statement - Greenbrier River crested at Buckeye and Marlington 4am
29	7:30 am	"	"	River Flood Statement - Cheat river crested at Rowlesburg, crest unknown 7am
30	8:00 am	"	"	River Flood Statement - Gauley River crested at Camden. River flood warning terminated....No further flooding expected on the Gauley River 7 am
31	8:50 am	"	"	River FLood Statement - West Fork river at crest - Clarksburg 8 am
32	11:15 am	"	"	River FLood Statement - Little Kanawha crested at Glenville at 7am
33	12:30 pm	"	"	River Flood Warning - for Monogahelia River
34	2:30 pm	"	"	River Flood Statement - REcord flooding cont. ofr Greenbrier & Little Kan.,
35	5:42 pm	"	"	River FLood Warning - West Fork River, valve failed at Stonewall Jackson Dam
36	7:35 pm	"	"	River Flood Warning - West Fork River revision - valve has been placed under control so rapid rise is not expected
37	9:15 pm	"	"	River Flood Warning - Tygart Valley river crested at Dailey, Belington and Philippi.
38	12:45 am	Nov. 6	Charleston	River Flood Statement - Mon River crested at Morgantown at 3 pm at 27'
39	1:30 am	"	"	River FLood Statement - Greenbrier crested at Alderson at 11 am at 24.5
40	4:45 am	"	"	Bulletin - Special River Statement Correction - WV rivers Break Records This bulletin gives all stats.
41	10:00 am	"	"	River Flood Statement - Little Kanawha crested at Creston at 7am at 28.07'
42	10:35 am	"	"	River Flood Statement - Major flooding cont. at Philippi on Tygart Valley River
48	11:10 am	"	"	Flood Warning - Ohio River T R Lock & Dam expected to crest at 3:30 pm
43	4:30 pm	"	"	River Flood Statement - W. Va. River Summary
44	10:00 am	Nov. 7	Charleston	Flood Statement - Rivers cont. to recede in aftermath of record flooding
45	10:40 am	"	"	Flood Statement Correction - Rivers cont. to recede in...
49	3:45 pm	"	"	Flood Statement - Ohio River near crest a Racine Lock & Dam
50	6:45 pm	"	"	Flood Statement - Ohio River crested at Racine Lock & Dam at 39.3'

Bord, Harman take supplies into flood ravaged counties

Nov. 9
Mountain Statesman
Grafton
By Bob Bucy

Like walking into hell—that's how local residents John Bord and Tom Harman described their Wedsnesday night excursion into the flood-stricken areas of North-Central West Virginia.

The city began organizing its flood relief campaign on Wednesday and early that evening, Bord and Harman decided to make the trek to Parsons and Rowlesburg. They headed out in a two-wheel drive pickup truck, that was loaded to the brim with clothes, food and other relief items. "Tom said that afternoon, let's go, so around 7 p.m. we left." Bord said. "We didn't have any idea what we'd be getting into and we had no idea what we would see," Bord said.

What they did see will stay with them for a lifetime.

They first headed to Parsons and one scene caught their eye as they entered that town. "There was a mobile home wrapped around a telephone pole as we went in. It was incredible. Another mobile home was shredded apart and a man was inside walking around," Harman said. "As we got on into Parsons it just looked like someone had poured water over top of it and then started stirring. Cars were turned over on top of cars—I'd almost be afraid to go back," Harman said.

"The people in Parsons told us about St. George, which is about four miles away," Bord said. "They were very concerned about the people in St. George and they wanted to share all the supplies they could with St. George." St. George is a small community of a few hundred people, which had been isolated from the outside world following the flood. Bord and Harman headed for the town, but a major obstacle got in the way.

"There's a bridge to cross, before getting into St. George, and the people in Parsons weren't sure if it was open. We came up to it and it was pitch black outside. There was a cord hanging down, as if the bridge was closed," Bord explained. "We had to make a decision and John said, 'Let's go, God is with us,' and we went across it," Harman said. (The bridge is around half the length of the St. Mary's bridge.)

"Of all times I've ever been afraid, it couldn't equal what I went through going across that bridge, wondering if it would hold and knowing how swift the current was below. It's something I'll never forget."

Once over the bridge, staring into the darkness, Bord realized that the road had been washed out. The National Guard had plowed a road through a field. "It was all mud. It's amazing we made it through with a two-wheel drive," Bord said.

Upon reaching St. Geroge, Harman and Bord realized that they were the first to provide aid to this community. "It was a pitiful site. Some people were packed into a picnic type shelter that had concrete floors and they had a stove that was almost out of fuel," Bord said. "The people there were just so glad to see us. One man said, 'You came all the way from Grafton to help us?' they were just overjoyed to see us.

Next, the pair headed to Rowlesburg, which had an 8 p.m. curfew. However, the National Guard allowed Bord and Harman entrance since they had supplies. "It was like driving into a nightmare," Bord said. "As soon as we got in, Tom said Hell can't be this bad and I had to agree. You cannot fully know the effects of the destruction in a place like Rowlesburg until you see it. There was absolutely no traces of some houses."

"Everything people talked about was in the past tense," Harman said. "They had been stripped of their pride, dignity and everything they owned and it was like a brand new beginning. They have to start all over." Harman said that one new home had just been finished. After the flood, a few layers of bricks were all that remained.

Bord said firemen in the town had worked for 36 hours straight.

BELINGTON—This Barbour County community was devastated by floodwaters which reportedly destroyed 43 businesses and damaged 85 homes. (Courtesy of The Democrat)

"One man parked his jeep outside a house to rescue the man inside. He went in, carried the man out and found his jeep under water. He waded out carrying the man," Harman said. "There's a lot of heroes in Rowlesburg," Borg said. "And in all those towns."

The pair returned home around 1:30 a.m. Thursday and have since made other trips to distribute supplies. "I haven't seen one person cry yet," Harman said. "I can't understand that. I guess they're just stunned. Maybe the full effects won't hit them until they try and pull things together."

Bord explained his decision to cross the treacherous bridge. "I knew we had things that people on the other side needed. We couldn't be concerned with not making it. Then, we drove right through the mud, without ever getting hung up. I don't think there's any question that someone was looking over us."

Paw Paw concerned bridge added to flooding

Nov. 20
The Morgan Messenger
Berkeley Springs

Paw Paw and Morgan County officials met twice last week to discuss whether the design of the new Paw Paw bridge raised water levels in the town and contributed to flooding problems on November 5.

Before work began to replace the bridge, the approach to the structure was open underneath, allowing river flow when the Potomac overran its banks. Before the old bridge was taken down last month and a temporary bridge put in use, about 26,500 cubic yards of fill was put under the approach. The fill reduced the width of the bridge from 640' to 444', it is claimed.

Officials of Paw Paw Industries and Consolidated Orchards, the town's major employers, wonder if some of their flooding was caused by the fill holding back and rerouting water.

Both Paw Paw Industries and Consolidated Orchards have told Paw Paw Mayor David Clark that they are reluctant to sink money into their flooded businesses unless the problem can be corrected.

Floodwaters in Paw Paw rose 54 ft., about the same as in 1936. Flood stage is considered 25 ft. Downstream areas generally received a few less feet of floodwater than in 1936.

Between them, Paw Paw Industries and Consolidated Orchards suffered about $9 million in flood damage. Neither had flood insurance. Both will require low interest government loans if they decide to reopen.

West Virginians reach out and help someone

Nov. 16
Parkersburg Sentinel
Parkersburg
By Dave Owen

Harold Kessel went to Petersburg to survey watershed control dams. He went back two days later to shovel mud.

Ken Calebaugh drove a truck load of supplies to Philippi. He's back there today, leading a work party of 20 volunteers.

Roberta Cox was marooned in Clarksburg when the flood struck. She's back in Clarksburg as part of the Red Cross emergency aid team.

Bob Hopper was working to get the Salvation Army's new facilities open. He's still here, but also running a major relief forwarding depot.

Phil Zebly had to take a motel room 20 miles from home when floodwaters blocked the highways. Today he's recruiting volunteer interviewers and cleanup crews and is into other relief efforts.

Nov. 4, floodwaters which exceeded the "100 year level" inundated more than half of West Virginia, the worst disaster in years, setting in motion a major search and rescue operation which is still under way.

Harold, Ken, Phil, Bobbie and Bob are typical of the caring people who volunteered to help and will be helping in the weeks ahead.

Kessell, area soil conservationist, was on a SCS team that made on-site inspections of the watershed flood control dams in Eastern West Virginia. "The dams all held, all did their job . . . I'm from that part of the state . . . the damage is unbelievable." He returned to the area, taking time from work, an almost instantaneous response to what he saw.

Hopper, comanding officer of the Salvation Army corps here, had officers and volunteers in the field as the flood waters crested . . . and the Parkersburg volunteers are still there, working along side other volunteers and Salvation Army officers and personnel from seven other states.

Canteens (fully equipped mobile kitchens) from Cincinnati and Columbus were dispatched immediately to the flood zone and are still there. The Salvation Army is serving thousands of meals a day. The Parkersburg volunteers are headed by Lt. Jackie Kelley and include Sgt. Major Paul and Jo Baker, Amanda Riddlebaugh, Diane Bowling, Quentin Beha, Jim Gordon and Edgar and Ruth Kent.

"Parkersburg is the central forwarding point for relief aid," Hopper said. "Sister Christian Krause of the St. Francis Relief Center in Vinton, Ohio, sent a truck load of materials and local collections are being delivered here. The other day Mitch Belt furnished the rig which took the materials to the Army's distribution center at the Barbour County Fairgrounds."

Coordinating Salvation Army efforts statewide is Major Stanley Jaynes, divisional secretary from Baltimore. He is one of the more than a score of out-of-state officers working in West Virginia.

Calebaugh is pastor of the Williamstown and Waverly Presbyterian churches. He recruited the work party in the two congregations and they will be working through the Heart and Hand Ministry at Philippi on clean up of homes and facilities such as the Senior Citizens Center. Earlier the youth of St. Mary's church went to Parsons to dig in and help out.

Cox, executive of the Mid-Ohio Valley Red Cross chapter, was visiting in Clarksburg and when she got back to Parkersburg, she coordinated the local relief aid and money efforts and worked on the statewide responsibilities shouldered by the Red Cross in this emergency. Now she's on a three-week assignment in Clarksburg, interviewing flood aid applicants.

Zebly, executive of the Presbyteries of Parkersburg and Grafton, is literally on the phone all the time—he's lining up work parties and interviewers to work with the Red Cross and Federal Emergency Aid Administration; he's involved in the Mountaineer Food Bank which distributed 152,000 pounds of food and he's bringing together the efforts of the congregations.

There's a common thread to these examples of the thousands who are helping—a care for our fellow man . . . that's the spirit that makes West Virginia great. Mops, shovels, buckets and paint alone can not repair the losses such as homes destroyed: 1,475 in Grant County, 134 in Calhoun, 21 in Gilmer and 33 in Tyler.

Telethons and flood aid concerts will raise money for immediate relief. Grocery firms are providing food at cost and free food. Broughton and Valley Bell dairies are bottling drinking water. These meet the pressing immediate needs. The first replacement housing is being brought into the state, Congress may lift the cap on emergency highway aid, but the people of West Virginia can not expect the federal government to fix the bridges, rebuild the roads, reopen the schools. It's going to take state investment too.

The Legislature is reviewing the situation and is willing to sit in special session to act on flood relief measures. And the Legislature should be willing, in an election year, to levy a special sales tax or a surtax on personal income taxes to finance flood repair. It must be done, the state must rebuild as the people rebound.

The material loss is in the millions of dollars; you can not place a dollar figure on the human loss. It's a heart rending situation; the response is heart-warming--West Virginians, fiercely independent by tradition, quickly moving to help one another as Bob, Harold, Phil, Ken, Bobbie and the Soldiers and Sallies of the Army have done.

PHILIPPI—Many residents were rescued by volunteer firemen and emergency squad members. Here, the rescue craft became temporarily snagged when the swift current carried the boat around the side of a pole opposite the mooring line. (Courtesy The Democrat)

PHILIPPI—From City Hall (right) to all points of the city, residents struggled to recover from the mire left behind by the Tygart Valley River. (Courtesy of The Democrat)

PARSONS—James Judy looks at one of his outbuildings that was overturned by the raging floodwaters. (Courtesy of Dominion Post)

'It will be a struggle back,' Rockefeller says

Nov. 20
The Morgan Messenger
Berkeley Springs
By Linda Swaim Buzzerd

Senator Jay Rockefeller was in Paw Paw Saturday afternoon to talk with homeless families and assess flood damages.

Rockefeller, accompanied by three members of his staff and two of his children, spoke briefly in Town Hall to a gathering of homeless residents and business owners hurt by the flood. He outlined steps to take to receive federal assistance. He asked to speak privately with each family and business owner to clarify specific needs and losses.

Two Homeless Familes

With two of the families' permission and the Senator's, this reporter sat in on their sessions. Lane Bailey, state director of Rockefeller's office, Phil McGance, transition officer, and Mike Robinson, Eastern Panhandle coordinator, took notes as the families talked and asked questions.

Jimmy and Teresa Campbell have three daughters, Amanda, 10, Liza, 7, and Natalie, 3. They lost their mobile home and all personal possessions.

Rockefeller advised them to document losses with pictures and records. The Campbells had been to the Federal Emergency Management Agency (FEMA) office in Romney to file for assistance. He advised them to file an amended 1984 tax return for flood losses.

The Campbells are living in a crowded situation. Rockefeller encouraged them to "be patient." He said it will take several months for the thousands and thousands of applications to get through federal computers in Atlanta.

It may be months, he said, until temporary housing (mobile homes) arrive in Paw Paw.

"We'll stand shoulder to shoulder and push through red tape. You'll get help. There is light at the end of the tunnel," he promised. He said the light may not be seen this week, or next month. "It will happen soon but not fast enough," he said.

Rockefeller said there are thousands of mobile homes thoughout the United States under contract to the government. Moving them into West Virginia's disaster area will take time. "Nobody knows yet how many we can get or where they come from," he said.

"Together, we'll stand with you," he told the Campbells. Rockefeller promised to get some questions answered and phone them this week.

The Fletchers

Anthony (Tony) Fletcher and wife, Sandra (Susie), have two children, Tony, Jr., 14, and Denise, 12. The Fletchers lost their mobile home and are living with his grandmother, Kathryn Avey, in Paw Paw.

Susie Fletcher told Rockefeller that she had also been to the FEMA office in Romney. She was told her family **did not make enough money** to qualify for a low interest SBA loan. And, she did, she was told they **did not qualify** for the entire $5,000 federal grant.

Taking Tony Fletcher by the shoulders, Rockefeller said, "It'll take the big broad shoulders you got. It will be a struggle back."

Rockefeller toured damaged areas in Paw Paw just before dark in heavy rain and left for Romney and areas west, three hours behind schedule.

Flood relief fund to benefit newspapers

From WVPA

West Virginia Press Association announced last Tuesday that a special relief fund had been established to benefit those newspapers in West Virginia which suffered extensive damage during last week's flooding.

Two weekly newspapers—the *Pocahontas Times* in Marlinton and the *Grant County Press* in Petersburg—lost all or most of their production equipment, along with fixtures and supplies, according to W.F. Childress, executive director of the Press Association.

Four other weekly newspapers—*The Parsons Advocate,* the *Moorefield Examiner,* the *Moorefield Democrat* and the *Glenville Democrat*—suffered varying degrees of flood damage, he said.

With the assistance of other newspapers and through "extraordinary" arrangements, all of the flood-affected nespapers have resumed publication.

Deaths

The Pocahontas Times
Marlinton

Larry and Jerry Regester, 34, identical twins of Garnet and Adeline Ricottilli Regester, of Moorefield, died during the flood on Monday, November 4, 1985. They had been on Cheat Mountain on a logging job and were on their way home when caught in flood water at the Smoke Hole Cavern. Four others died at the same location.

Funeral arrangements are incomplete.

They were nephews of Mr. and Mrs. Tony Ricottilli, of Marlington.

PARSONS—Ruined shoes from Parsons Footwear. (Courtesy of Elkins Inter-Mountain)

PARSONS—Parsons Footwear employees clean out the tons of mud left by the flood. (Courtesy of the Inter-Mountain)

HENDRICKS—Houses once lined this area which used to be Carl Alley. (Photo by John Warner)

The town of Hendricks is done for . . .

Nov. 9
Inter-Mountain
Elkins
By Nancy Isner

"The town of Hendricks is done for. The townspeople are walking around in a daze. They are helping each other as much as they can, but they are just dumbfounded," said Mabie resident Faye Bertelli who traveled to Hendricks this week to check on several relatives who lived in Hendricks.

Bertelli, who was raised in the small Tucker County town said, "I couldn't believe what I saw. The Hick Fairgrounds are completely gone. The old swinging bridge is gone. Charles Street next to the fairgrounds was completely destroyed," she said.

Bertelli said six persons were stranded on the second floor of the Russell Hebb house on Charles Street Monday night.

"As I understand it, there were two people in the house and four others went in to try and save them, but they all became stranded inside the house as the water began to rise.

"They had to go to the second floor because the water was gushing in the first floor. They all spent the night on the second floor of that house as water poured through it all night long. It was a miracle the house didn't collapse from all that water pouring in," she said.

"They had a flashlight and they kept flashing it to a nearby house to let those people know they were still in there."

When the water subsided the next morning the six stranded persons were able to leave the house unharmed, Bertelli said.

Residents also said two teenagers who serve as junior firemen who were going to the Hebb house to help rescue the persons became stranded by the flood waters and had to spend the night in a house across the street.

"There are two houses that are completely gone and there's no sign of them anywhere.

"One house was my cousin's, Peck Mitchell who lived near the fairgrounds. All he got out was the clothes on his back. The flood came that fast.

"Peck said he stood next to the railroad tracks and watched the water take his home off its foundation. Then a building came floating by and struck his house and knocked it down the river.

"Pieces of his house are now along the river at Hambleton," she said.

"He said he could hear the water coming. It came off the Black Fork and Dry Fork rivers with a gush and landed right in the bottom section of Hendricks," she said.

Bertelli said over 20 homes and trailers in Hendricks were either destroyed by Monday's flood or were left unliveable.

"So many of the houses left standing by the river can't be lived in and houses on both sides of Charles Street are all gone. They are all undermined and there's debris in them like you wouldn't believe. Near Charles Street in some homes the mud is six to eight feet deep."

Among the other properties destroyed was Mary Bowley's beauty shop, "but one amazing thing was the town's church," she said, which was located near the Hick Fairgrounds.

"All the church's chandeliers are completely intact and not a globe was broken. In fact there was a glass dish sitting on an inside window sash and that dish never budged," she said.

Bertelli said the church was bulging in the back and inside in the center of the church, "and the pews are everywhere inside."

She said members of her family and other persons entered the church to salvage what few items they could. "We removed a picture and also a plaque which was in memory of my brother, Robert Coontz, and a cross which was in memory of Isabelle Lindsey, along with several books.

"We wanted to get these things out so in case they did want to rebuild the church, these items could be returned."

↑ PARSONS—The approach to the Chessie Railroad trestle over the Black Fork River was wrenched apart by flood waters. (Photo by John Warner)

↓ PARSONS—Overturned vehicles were a common sight in downtown Parsons. (Photo by John Warner)

Can you spare $20 for a water-soaked book?

Nov. 20
News Leader
Richwood
By Jim Comstock

On a street in Parsons last week I saw enormous stacks of water-soaked cartons and realized that here was an irreplaceable flood loss which isn't apt to make the headlines. In these cartons, snugly packed, are new, unread, untouched books, all headed for a landfill instead of a book shelf.

This estimated million dollars loss represents the toils of a two-generation family engaged in producing, along with a weekly newspaper of the town, new and out-of-print West Virginia books. In these soggy cartons are copies of that Big Three of our State's beginning, hewed from the wilderness by our plucky ancestors, "Border Warfare" by Withers, "Notes on the Settlement and Indian Wars" by Joseph Doddridge, and "History of the Early Settlement" by Willis DeHass. And, too, there are new books, as new as last week in this ruined mess.

This is the awesome climax to an enterprise launched by the late Ken McClain and continued by his son-in-law George, and daughter Mariwyn Smith. This esteemed couple took up Ken McClain's torch of inspiration and held it high, adding title after title to that original list, giving new writers an outlet for their creativity. Now, it is all stacked up on the sidewalk, ready to be consigned to the earth with the great Ken McClain himself.

I saw this staggering period to endeavor one day last week with my newspaper partner and son Jay. We'd got this paper out, after having made our equipment and expertise available to help Jessee and BIll McNeel meet the deadline of her *Pocahontas Times,* which had been one of the bleak casualties of the Marlinton flood of last week, and we drove over Kennison Mountain and beyond to Parsons. There was good reason for this visit. Back in the beginning days of this paper, we suffered a backset, and Ken McClain, hearing of our troubles, made his *Advocate* equipment available to us. Without him we could easily have bellied up. We wanted to ask if forty years later, we could repay that debt.

We stopped on the way to view some of Marlinton's destruction, and as terrible as it was, it fared much better than Parsons. The destruction here is completely indescribable. Heartbreaking. But, the story has been well covered in the media, and I will therefore make no detailed comments.

Jay and I are printers and publishers enough to know the loss that George and Mariwyn Smith suffered at their *Advocate* quarters. A $100,000 press, recently purchased with a bank loan, and with a capability of turning out books in color, would never function again. Working men, and one girl, were cleaning mud away from the presses. But the big loss as we deemed it was the destruction of plates for the immediate replacement of sold-out books.

We didn't get to talk with the Smiths as they were in attendance at a mass meeting of the flood sufferers. However, we did learn there was nothing we could do to help with the paper, as they had made arrangements, we were told, to have their paper printed in Elkins, thirty miles away.

However, there is something we can do.

Briefly, let's not let George bury those water-soaked books in a landfill. Let's each of us send him $20 for one of those books, and put it on our book shelf as a souvenir to keep and remember by its presence what these people behind it stood for, and to hand down in the family. Odd, you might say, a souvenir of a man's unhappy loss, but that little pepping up gesture may sustain the Smiths in their bitter hours. Not to speak of the bonanza of a few thousand dollars in a starting fund again.

PARSONS—Sightseers were not welcome in St. George or any of the other flood damaged communities. (Courtesy of Grafton Mountain Statesman)

PARSONS—Many homes in Tucker County though still standing, would be condemned shortly thereafter. (Courtesy of the Inter-Mountain)

ST. GEORGE—A tree near this Tucker County community reveals the high water level during the flood. (Photo by John Warner)

Flood counties in profile

BARBOUR COUNTY

- ★ 30 Households Displaced
- ★ 4 Bridges Affected
- ★ 259 Private Homes Destroyed
- ★ 2 Private Homes Condemned
- ★ 139 Private Homes With Major Damage
- ★ 46 Private Homes With Minor Damage
- ★ 5 Private Businesses Destroyed
- ★ 119 Private Businesses With Major Damage
- ★ 2 Public Buildings With Major Damage

BERKLEY

- ★ 7 Households Displaced
- ★ 80 Private Homes Destroyed
- ★ 31 Private Homes Condemned
- ★ 46 Private Homes With Major Damage
- ★ 26 Private Homes With Minor Damage
- ★ 1 Road Affected

BRAXTON

- ★ 25 Households Displaced
- ★ 3 Private Homes With Major Damage
- ★ 56 Private Homes With Minor Damage
- ★ 4 Private Businesses With Minor Damage

CALHOUN

- ★ 30 Households Displaced
- ★ 134 Private Homes Destroyed
- ★ 38 Private Homes With Major Damage
- ★ 20 Private Homes With Minor Damage

DODDRIDGE

- ★ 31 Households Displaced
- ★ 16 Private Homes Destroyed
- ★ 17 Private Homes With Major Damage
- ★ 40 Private Homes With Minor Damage
- ★ 6 Private Businesses With Minor Damage

GILMER

- ★ 25 Households Displaced
- ★ 21 Private Homes Destroyed
- ★ 193 Private Homes With Major Damage
- ★ 60 Private Homes With Minor Damage
- ★ 1 Private Home Condemned
- ★ 2 Private Businesses Destroyed
- ★ 38 Private Businesses With Major Damage
- ★ 73 Private Businesses With Minor Damage
- ★ 1 Public Building Destroyed

GRANT

- ★ 500 Households Displaced
- ★ 1 Bridge Affected
- ★ 10 People Dead
- ★ 4 Missing
- ★ 3 Schools Affected
- ★ 1,475 Private Homes Destroyed
- ★ 146 Private Homes Condemned
- ★ 225 Private Homes With Major Damage
- ★ 150 Private Homes With Minor Damage
- ★ 36 Private Businesses Destroyed
- ★ 2 Private Businesses Condemned
- ★ 69 Private Businesses With Major Damage
- ★ 9 Private Businesses With Minor Damage
- ★ 2 Public Buildings With Major Damage

GREENBRIER

- ★ 25 Households Displaced
- ★ 1 Bridge Affected
- ★ 42 Private Homes Destroyed
- ★ 8 Private Homes Condemned
- ★ 192 Private Homes With Major Damage
- ★ 99 Private Homes With Minor Damage
- ★ 3 Private Businesses Destroyed
- ★ 2 Private Businesses Condemned
- ★ 8 Private Businesses With Major Damage
- ★ 29 Private Businesses With Minor Damage
- ★ 1 Public Building With Major Damage
- ★ 1 Public Building With Minor Damage

HAMPSHIRE

- ★ 40 Households Displaced
- ★ 1 Bridge Affected
- ★ 3 People Dead
- ★ 529 Private Homes Destroyed
- ★ 55 Private Homes Condemned
- ★ 39 Private Homes With Major Damage
- ★ 8 Private Homes With Minor Damage
- ★ 28 Private Businesses Destroyed

HARDY

- ★ 225 Households Displaced
- ★ 1 Bridge Affected
- ★ 1 Road Affected
- ★ 4 People Dead
- ★ 1 Missing
- ★ 600 Private Homes Destroyed
- ★ 120 Private Homes Condemned
- ★ 40 Private Homes With Major Damage
- ★ 10 Private Businesses Destroyed
- ★ 5 Private Businesses Condemned
- ★ 60 Private Businesses With Major Damage
- ★ 1 Public Building Condemned
- ★ 1 Public Building With Major Damage

HARRISON

- ★ 120 Households Displaced
- ★ 1 Bridge Affected
- ★ 56 Private Homes Destroyed
- ★ 11 Private Homes Condemned
- ★ 383 Private Homes With Major Damage
- ★ 171 Private Homes With Minor Damage
- ★ 6 Private Businesses Destroyed
- ★ 77 Private Businesses With Major Damage
- ★ 12 Private Businesses With Minor Damage
- ★ 1 Public Building Destroyed

JEFFERSON

- ★ 30 Households Displaced
- ★ 46 Private Homes Destroyed
- ★ 25 Private Homes Condemned
- ★ 25 Private Homes With Major Damage
- ★ 23 Private Homes With Minor
- ★ 1 Private Business Destroyed
- ★ 1 Private Business Condemned
- ★ 2 Private Businesses With Minor Damage
- ★ 2 Public Buildings Destroyed
- ★ 18 Public Buildings With Major Damage

LEWIS

- ★ 49 Households Displaced
- ★ 4 Private Homes Destroyed
- ★ 2 Private Homes Condemned
- ★ 158 Private Homes With Major Damage
- ★ 109 Private Homes With Minor Damage
- ★ 3 Private Businesses With Major Damage
- ★ 15 Private Businesses With Minor Damage
- ★ 1 Public Building With Major Damage

MARION

- ★ 1 Private Home Destroyed
- ★ 29 Private Homes With Major Damage
- ★ 95 Private Homes With Minor Damage
- ★ 7 Private Businesses With Major Damage
- ★ 2 Private Businesses With Minor Damage

MINERAL

- ★ 1 Bridge Affected
- ★ 4 Private Homes Destroyed
- ★ 52 Private Homes With Major Damage
- ★ 74 Private Homes With Minor Damage
- ★ 5 Private Businesses With Minor Damage

MONONGALIA

- ★ 100 Households Displaced
- ★ 102 Private Homes With Major Damage
- ★ 2 Private Businesses With Major Damage
- ★ 4 Private Businesses With Minor Damage

MONROE

- ★ 25 Households Displaced
- ★ 1 Private Home Destroyed
- ★ 26 Private Homes With Major Damage
- ★ 32 Private Homes With Minor Damage

MORGAN

- ★ 15 Households Displaced
- ★ 27 Private Homes Destroyed
- ★ 20 Private Homes Condemned
- ★ 21 Private Homes With Major Damage
- ★ 16 Private Homes With Minor Damage
- ★ 11 Private Businesses With Major Damage

"It was just unreal. I couldn't believe it happened. The water came so fast you couldn't do anything about it. You couldn't save anything. It was like a nightmare."

—— Craig Hare, flood survivor

"The lights started going off and the fire siren went off, and everybody got panicky and started screaming and yelling."

—— Eric Neilson, flood survivor

"It's funny, the stuff that stays close, even though your whole house washes away."

—— Allen Myers, flood survivor

"We moved here from California to get out of the earthquakes . . ."

—— Edith McDowell, flood survivor

NICHOLAS

★ 13 Private Homes With Minor Damage
★ 3 Private Businesses With Minor Damage

PENDLETON

★ 380 Households Displaced
★ 20 Bridges Affected
★ 9 Roads Affected
★ 10 People Dead
★ 4 Missing
★ 7 Schools Affected
★ 303 Private Homes Destroyed
★ 119 Private Homes Condemned
★ 163 Private Homes With Major Damage
★ 185 Private Homes With Minor Damage
★ 5 Private Businesses Destroyed
★ 1 Private Business Condemned
★ 8 Private Businesses With Major Damage
★ 15 Private Businesses With Minor Damage
★ 3 Public Buildings Condemned
★ 4 Public Buildings With Major Damage
★ 1 Public Building With Minor Damage

POCAHONTAS

★ 70 Households Displaced
★ 6 Bridges Affected
★ 5 People Dead
★ 115 Private Homes Destroyed
★ 43 Private Homes Condemned
★ 311 Private Homes With Major Damage
★ 39 Private Homes With Minor Damage
★ 32 Private Businesses With Major Damage
★ 38 Private Businesses With Minor Damage
★ 2 Public Buildings Destroyed
★ 1 Public Building Condemned
★ 20 Public Buildings With Major Damage

PRESTON

★ 346 Households Displaced
★ 7 Bridge Affected
★ 4 Roads Affected
★ 2 Dead
★ 2 Schools Affected
★ 204 Private Homes Destroyed
★ 70 Private Homes Condemned
★ 20 Private Homes With Major Damage
★ 2 Private Homes With Minor Damage
★ 15 Private Businesses Destroyed
★ 3 Private Businesses Condemned
★ 15 Private Businesses With Major Damage
★ 5 Private Businesses With Minor Damage
★ 4 Public Buildings Destroyed
★ 3 Public Buildings Condemned
★ 3 Public Buildings With Major Damage
★ 1 Public Building With Minor Damage

RANDOLPH

★ 50 Households Displaced
★ 9 Bridges Affected
★ 3 Roads Affected
★ 146 Private Homes Destroyed
★ 13 Private Homes Condemned
★ 239 Private Homes With Major Damage
★ 80 Private Homes With Minor Damage
★ 17 Private Businesses With Minor Damage
★ 2 Public Buildings Condemned
★ 1 Public Building With Minor Damage

SUMMERS

★ 50 Households Displaced
★ 1 Bridge Affected
★ 21 Private Homes Destroyed
★ 3 Private Homes Condemned
★ 23 Private Homes With Major Damage
★ 98 Private Homes With Minor Damage
★ 8 Private Businesses With Minor Damage
★ 10 Public Buildings With Major Damage

TAYLOR

★ 7 Households Displaced
★ 1 Private Home Destroyed
★ 4 Private Homes With Major Damage
★ 51 Private Homes With Minor Damage
★ 1 Private Business With Minor Damage
★ 1 Public Building With Minor Damage

TUCKER

★ 352 Households Displaced
★ 3 Bridges Affected
★ 2 People Dead
★ 1 Missing
★ 3 Schools Affected
★ 275 Private Homes Destroyed
★ 92 Private Homes Condemned
★ 266 Private Homes With Major Damage
★ 28 Private Homes With Minor Damage
★ 19 Private Businesses Destroyed
★ 9 Private Businesses Condemned
★ 22 Private Businesses With Major Damage
★ 15 Private Businesses With Minor Damage
★ 50 Public Buildings Destroyed
★ 7 Public Buildings With Major Damage

TYLER

★ 10 Households Displaced
★ 20 Private Homes Destroyed
★ 23 Private Homes With Major Damage
★ 18 Private Homes With Minor Damage
★ 2 Private Businesses With Minor Damage

UPSHUR

★ 25 Households Displaced
★ 2 People Dead
★ 3 Private Homes Destroyed
★ 225 Private Homes With Major Damage
★ 130 Private Homes With Minor Damage
★ 3 Private Businesses With Major Damage
★ 20 Private Businesses With Minor Damage
★ 1 Public Building With Major Damage
★ 1 Public Building With Minor Damage

WEBSTER

★ 20 Households Displaced
★ 1 Road Affected
★ 6 Private Homes Destroyed
★ 1 Private Home Condemned
★ 13 Private Homes With Major Damage
★ 53 Private Homes With Minor Damage
★ 1 Private Business Destroyed
★ 10 Private Businesses With Minor Damage
★ 1 Public Building With Minor Damage

Glenville people miffed, miserable

Nov. 8
Herald Dispatch
Huntington
By Dave Peyton

Residents of this Gilmer County town, hit hard by the flood earlier this week, were wondering yesterday if somehow the flood might have washed them off the face of the earth and out of everyone's memory.

Until yesterday, they had received little indication that anyone in the state knew about their plight.

They listened anxiously to radio and television reports to see if anyone took note that their downtown was virtually destroyed by the floodwaters from the Little Kanawha River.

"All we heard about was Weston. No one ever mentioned the problems we were having here," said Mack Samples, an official at Glenville State College. "We were beginning to wonder if we still existed."

Yesterday, however, the tragic story of Glenville began to unfold as people from the outside began to visit the town of 2,200 people and the county seat of Gilmer County.

What they saw, some said, was concentrated destruction as bad as any place in the state.

Mayor John Bennett had the tally by about noon yesterday. A total of 527 residents and 231 commercial buildings were flooded. Most of them had at least 6 feet of water. Scores of houses and apartments across the river suffered the same fate.

There was no doubt about it, they said. It was the worst flood Glenville has experienced in modern times.

For all of Gilmer County with its 8,800 people, the statistics were even more sobering. A total of 752 families were driven from their homes in the county and nearly 260 commercial buildings were inundated. Bennett said the figures may be higher once the reports are all in.

Most of those affected said they felt forgotten by the state.

"We don't amount to anything until they want our tax money," said Joan Layne, who was helping with cleanup efforts at her daughter's shoe store on Glenville's main street.

State Delegate Marjorie Burke, D-Gilmer, echoed those sentiments. "We've been left out. Our business district had been totally wiped out. The damage is much worse than at Weston (about 30 miles away). A lot of the merchants there weren't flooded.

"And I know things are bad in Clarksburg. But at least the people in Clarksburg have a few grocery stores still open. We had three grocery stores here and they're all flooded."

As an example of how Glenville and Gilmer County have been ignored, Delegate Burke noted that, "We've run out of tetanus vaccine and they couldn't send us any. A boy who's a native of Glenville but living in Charleston volunteered to deliver it to us in his car."

To make matters worse, she said, no media had showed up in Glenville until yesterday. When a camera crew from the Weston television station showed up, Delegate Burke chastised them for not arriving earlier.

Two marks outside the door of CJ Sports Inc. told the story of the tragedy. The first mark was a green line about 4 feet above the sidewalk. It marked the height of the 1967 flood. Two feet above it, scribbled in ink was the mark made for the flood that crested Tuesday morning.

"It was supposed to be the very last flood Glenville ever had," said John Jamison, owner of CJ Sports, pointing to the 1967 flood mark. "Burnsville Dam was supposed to stop all the flooding."

Burnsville Dam, 17 miles above Glenville on the Little Kanawha River, was completed in the early 1970s. And there were no more major floods in Glenville after that until this week.

"People were lulled into a false sense of security," Jamison said. "No one really believed we'd ever have a flood to match '67, much less 2 feet higher," he said.

MOOREFIELD—Mobile homes were smashed up against houses by the force of the floodwaters. (Courtesy of The Moorefield Examiner)

MOOREFIELD—Amidst the rubble left by the flood, an American flag was a cheerful sight on Veterans Day. Larry Heavner, commander of American Legion Post 64 distributes flags that line the streets. (Courtesy of The Moorefield Examiner)

MOOREFIELD—Drinking water was trucked in to Moorefield from as far away as Winchester, VA. (Courtesy of The Moorefield Examiner)

MOOREFIELD—The flooded intersection of Main Street and Winchester Avenue in Moorefield. (Courtesy of The Moorefield Examiner)

MOOREFIELD—A volunteer shovels out the mud from a Moorefield Elementary School classroom. (Courtesy of The Moorefield Examiner)

MOOREFIELD—Colt's Restaurant is marked with a B-2 (left side of picture)—condemned to be demolished. (Courtesy of The Moorefield Examiner)

PETERSBURG—A National Guardsman carries a child while escorting a resident from the ruins of a neighborhood. (Courtesy of News Tribune)

PETERSBURG—The downtown area resembles a war zone after the flood as MP's and National Guardsmen move in to aid recovery efforts. (Courtesy of the Inter-Mountain)

PETERSBURG—Many business properties like this one were not easily recognized when the waters receded. (Courtesy of Hampshire Review)

PETERSBURG—Floodwaters smashed this car through the wall of the Monarch Food Market in the southern part of town when a levy gave way. (Courtesy of News Tribune)

PETERSBURG—Heavy equipment was a common sight as workers fought to open streets and roadways. (Courtesy of News Tribune)

PETERSBURG—Residents and military personnel begin assessing damages after streets are cleared of mud and debris. (Courtesy of News Tribune)

Flood waters inundate Keyser, area

Nov. 5
Mineral County News Tribune
Keyser
By Mona Ridder

Sirens screamed intermittently throughout the night and the streets of Keyser resembled a scene from the fertile imagination of a science fiction writer bent on disaster as rising flood waters inundated the roadways and virtually cut the city off except for the Green Mountain Road and Memorial Bridge last night. City residents moved about the streets, mostly on foot, many fleeing to higher ground and others just curious, watching the water in New Creek as it relentlessly crept even higher, finally spilling over its banks, damaging cars, homes, and businesses.

Roadways throughout Grant, Mineral, Hampshire, Garrett, and Allegany counties were flooded cutting off residents and stranding travelers.

U.S. Route 220 was closed at the Potomac Motel and at Samson's Restaurant at approximately 8:30 p.m. when water reached depths that would not allow even 4-wheel-drive vehicles to pass. Route 46 at the corner of Armstrong and Water Streets was closed about the same time and water at that site eventually reached the windows of the Magistrate Court building before beginning to recede about 1 a.m., when the recorded rainfall for the previous 24 hours was set at seven to eight inches for the area.

PETERSBURG—Even some buildings which remained standing after the flood were later marked for demolition by officials as evidenced by the "A-113" spray-painted on this structure. (Courtesy of Piedmont Herald)

'. . . I was too scared to stop'

Nov.9
The Intelligencer
Wheeling
By Warren Bays

Crawling on her hands and knees straight up the side of a mountain, 71 year old Mildred Ernest of Triadelphia was afraid to look back at Smoke Hole Lodge. She could hear the water coming and through the darkness of a rainy night she prayed her way toward the safety of high ground.

She didn't know how high she should climb because the roaring grew louder and soon the cracking of wood echoed through the night. Smoke Hole Lodge, the nearly 100-year-old Grant County mountain retreat owned by the Stifel family of Wheeling, fell like a house made of Tinker Toys under the weight of a mountain of water.

On Monday night, climbing a mountainside in the dark to get away from that raging river, Mildred Ernest found her way to a hollow tree where she spent the night listening to the worst flood in West Virginia's history as it changed the face of the mountain-land forever.

Ernest had been a cook at Smoke Hole Lodge since September. She left the lodge Monday night with 19 year old Robby Gallagher of Wellsburg to seek safety from the storm. Stifel was trapped on the opposite side of the river from the lodge and yelled to Ernest that he would walk down the mountainside toward the town of Petersburg and call for help.

"I wasn't thinking of anything but climbing. I grabbed from one tree to another," Ernest said Friday in Wheeling. "Robby was off in another direction and at one time I slid down about 25 or 30 feet and was stopped by a tree. But I was too scared to stop. I could hear the lodge breaking up. When Robby came back over to me, we both huddled together, shaking like wet leaves, inside that hollow tree."

On Thursday, Stifel's family rented a helicopter to begin a search for him. Jack Waterhouse flew over Smoke Hole and was the first

PETERSBURG—An aerial view reveals that floodwaters remained trapped in and around the properties in the downtown area. (Courtesy of News Tribune)

family member to learn the lodge had been destroyed. Stifel, Ernest and Gallagher, who tends to cattle on Stifel's Smoke Hole retreat, were later located. Family members said those who lived through a horrifying experience "still are in a state of shock. But they'll be OK. Eddy wants to stay in Petersburg. Smoke Hole meant the world to him," his mother said Thursday night.

The Eastern Panhandle, particularly Grant and Hardy counties, have become popular second homes for hundreds of Northern Panhandle residents, many of whom expressed a homeless feeling this week because of the lack of news emanating out of the two counties.

But the attractiveness of the region to urbanites is the same reason for a lack of news during a disaster. The area is remote. "The wind is your neighbor," the farmer says, and word travels slow."

When telephone service was restored Friday afternoon, the picture described by residents of the area was ugly.

"You drive through Moorefield and it makes you want to cry," Vetter said. "It didn't miss anybody, this flood. There are old houses and new houses just washed off their foundation. Many were thrown about and others just buckled and bent. Trailer courts have disappeared from the middle of town. Mobile homes are piled on top of each other."

The human suffering is enormous.

"The residents are receiving two gallons of drinking water each day. That's all there is," Vetter said. "The water's being trucked in from Winchester, Va. The National Guard is posted throughout both counties and you don't move about much after 5 p.m. It's just one hell of a big mess. What food and water there is seems to be coming in by trucks, but they hope to have the Foodland store reopened by Monday.

"It's too hard to describe. But you walk through the middle of town and see buildings just destroyed. Outside of town, you won't recognize the country," Vetter continued. "Streams became rivers and when they flooded they changed the looks of this country. Apartment houses are gone from the town. Hunting lodges are gone from the country. Camping areas are destroyed. It just seems like the best thing to do would be dig a big hole and bury the debris and start all over rebuilding the community."

But men the mountains made have already started reshaping their lives.

"There are temporary roads being built and people are setting about cleaning up what is left. We'll do whatever is necessary to get back to normal. But this week really changed this country. It's left some pretty bad scars," Vetter added.

When Moore and a host of state and federal officials flew over the devastation earlier in the week, the governor said it was the worst scene he had ever witnessed. He called the flooding and ruination of whole communities West Virginia's worst disaster.

When Waterhouse landed in a helicopter in Petersburg on Thursday, he was asked, "Have all of you forgotten about us?"

Waterhouse said the destruction of entire communities is "hard to imagine until you actually see it. Then, it makes you sick."

He added, "It's even worse to feel like you are alone at a time like this. Those people think the rest of the state and the country has forgotten about them. I wish they could know we haven't forgotten them and we do care."

Waterhouse, Lloyd Walker and Joe Taylor will lead a caravan of vehicles to Grant County this morning with food supplies for people of the Petersburg area.

Residents in other Northern Panhandle communities have made similar plans.

Officials have urged anyone able to help the victims of the flooding to contact their local Red Cross chapter or the Salvation Army. Everything imaginable, from water to clothes, medicine to furniture is needed for victims in the Eastern Panhandle area.

→

HAMPSHIRE COUNTY—The railroad bridge and approacn parallel to U.S. Rt. 50 were swept away by powerful currents. (Courtesy of News Tribune)

PETERSBURG—Twisted trees and grim shards of concrete sidewalks mark what was once a busy city block in the Grant County seat. (Courtesy of Grant County Press)

THE GAP—The familiar Petersburg Gap Bridge disappeared when the Killing Waters passed through. (Courtesy of Hampshire Review)

Pleasants Countians tell tales of flood's havoc

Nov. 14
St. Marys Oracle
St. Marys

In the aftermath of heavy rain and severe flooding throughout sections of West Virginia last week, the stories keep coming in of families who have relatives here and who were in the path of the runaway rivers and streams.

Pleasants County's damages were slight, compared to such neighboring towns as Glenville, where the Kanawha River swept through downtown streets.

One local family's relatives were in the path of the deluge. Lorene Cunningham said Monday that she had "never seen anything like it," although her parents, Albert and Ruth Ellyson, have lived in the Glenville area for 50 years.

Flood waters inundated the first floor of the Ellyson's two-story home, lacking about 18 inches from going into the upstairs, Mrs. Cunningham said. Her parents had been living on the first floor of their home, not using the second floor.

Luckily, they were not at home at the time of the flood, but were in Charleston visiting a daughter. Another daughter, from Canton, brought a vacuum machine over and cleaned the water out of the carpets this week, while the Cunninghams helped gather household furnishings and restore the house to a semblance of order.

Other members of her family affected by the flood included a brother, Gene, and his wife, Edith, who lost four tractors, along with all the hay and corn to feed his stock through the winter.

In addition, flood waters inundated their house, with 16 inches of water on the second floor porch, and the water came up over it, too.

Flood victims pick up pieces

Nov. 7
Parkersburg News
Parkersburg
By Andrea Novel

Gilmer County residents Wednesday began a massive cleanup from heavy floodwaters that forced 1,000 people from their homes and destroyed downtown businesses in Glenville.

The floodwaters, measuring as high as 10 feet in some places, had receded by 5 a.m. Wednesday, leaving the town drenched and mud-spattered. A volunteer cleanup crew followed the receding waters, bulldozing out damaged store goods and hosing out buildings and homes.

Residences and businesses in downtown Glenville, Brooklyn, Sand Fork and Hays City areas all had been under water since Monday, when heavy rains pummeled the north central and eastern parts of the state, causing extensive flooding and 18 deaths in West Virginia.

No fatalities and only ne near drowning were reported in Gilmer County during the disaster. The county was left without potable water and gas in some areas.

West Virginia State Trooper First Class L.G. Smith said the hardest hit areas in the county were downtown Glenville, Brooklyn and Sand Fork. Some 95 percent of the town's businesses lost everything, he reported.

"We're really in pretty good shape for what we've went through," he said. He reported all primary and most secondary roads are now open in the county.

Kanawha Union Bank President Tom McPherson said the flood will be hard on the community, especially for the people who don't have flood insurance.

"We're not unlike a lot of small towns where business people are having difficulty," he said about the town's economic situation before the flood. "For some, it probably will be impossible to go back into business."

The bank experienced about 5 1/2 feet of water in its bookkeeping area and 16 inches in the building's main lobby. "Our situation here is mild compared to a lot of people's," McPherson said.

Donna Wyatt, co-owner of the Pioneer Grocery Store, said they only had time to get some milk, bread, meat and baby formula out of the store before it became inaccessible. "The water was at knee-high level when I left at 10:30 p.m. and at midnight it was up to the windows," she said. "That's how fast it was."

She and her business partners lost everything in both their grocery and its neighboring health and beauty aids store. All grocery stores in the town were under water including two in Hays City.

Ms. Wyatt expressed appreciation for area residents and Glenville State College students involved in the clean up. "You think we live in this fast world and nobody cares about anybody else," she said. "Friends, neighbors and people we didn't even know were here to help."

Brooklyn resident Bob Butcher, who has lived along the Little Kanawha River all his life, said the flood was the worst he has ever seen in the county. The flood of 1967 was the granddaddy of all floods, he said. "This one outdid the 1967 flood."

Forced to evacuate from his home by boat, Butcher questioned the value of the Burnsville Dam, built about halfway along the Little Kanawha River some six years ago. "The Burnsville Dam was supposed to be the flood control of the Little Kanawha River.

"We've got that big dam up there, and we have the greatest flood in history," he said.

Other Gilmer countians expressed similar concern about the dam. Some county residents had even dropped their flood insurance policies once the dam was built.

SPRINGFIELD—The water crested to the rooftops of many homes along the South Branch. (Courtesy of Hampshire Review)

SPRINGFIELD—The bridge remains standing but the approaches to it have been washed away at Millesons. (Courtesy of Hampshire Review)

ROMNEY—The South Branch's rain-swollen waters spilled into all low-lying areas. (Courtesy of Hampshire Review)

"The two opened the trailer door to find a virtual sea of mud and debris within. Small fish flopped helplessly in the sticky mixture."

—— Martinsburg Evening Journal

Second highest flood crests early Tuesday at 24.3 ft.

Nov. 6
Weston Democrat
Weston

The most devastating flood since the record flood of 1950 struck Weston and Lewis County Monday and early Tuesday morning when the West Fork River crested about 5 a.m. Tuesday at 24.3 feet, slightly less than a foot lower than the 25.5 foot 1950 crest.

No injuries or loss of life were reported but property damages are expected to run into the hundreds of thousands of dollars.

Ironically, the Stonewall Jackson dam lacked 62 feet in one of its sections from being completely top of dam. Water was being released through three flood control sluices Monday but late in the day it was reported that water was beginning to flow over the uncompleted monolith.

Business and other activities came to a standstill Tuesday morning with Weston isolated in four directions, including I-79 where both north and south bound lanes were closed near the Jane Lew exit. West Virginia Glass and Louie Glass were closed Tuesday as was the Greif Box Factory at Homewood.

Schools were dismissed early Monday and remained closed Tuesday.

Total rainfall up to 7 a.m. Tuesday was 4.95 inches according to Bill Woodburn, resident engineer, Stonewall Jackson lake project. At 1 p.m. Monday there had been a total of 4.5 inches since Monday morning. Woodburn said water began going over the top of the unfinished monolith number four Monday afternoon. This probably kept the river below the dam at its upper heights for an extended period.

An emergency evacuation center was set up at the Columbia club under the direction of the disaster committee of the local Red Cross chapter. George Whelan, chairman, reported approximately 50 persons were bedded down there Monday night on mattresses provided by the Weston hospital. Food was furnished from the Central school. Eighteen school children were caught at the school by the rapidly rising waters Monday and were taken to the Columbia club. A busload of children returning from a field trip to Charleston was forced to stop at Walkersville where they were housed at the fire department until they could return to Weston.

Bad as the flood was, it was well below the forecast issued by the National Weather bureau at Charleston which Monday continued to predict a flood crest for Weston of 27 feet by 10 p.m. This would have been a foot and a half above the flood of record.

Following heavy rains on Sunday the water came up rapidly Monday morning with high water reports at the fire department coming from West Second, Turnertown, Broad Street, Route 19 North, Gee Lick, Dry Fork, Kincheloe road, and at the bridge at Good Hope. At 8 a.m. Monday there had been 2.5 inches of rain.

At 4:45 p.m. Monday there was 21.7 feet of water at the gauge and at 5:55 p.m. it had risen to 22.03 and was pushing the 1967 mark. Tuesday night it reached a crest of 24.3 feet about 5 a.m. By 10 a.m., Tuesday, the water had dropped about four inches.

The National Guard was ordered into action Monday and three large vehicles were soon removing people from the worst areas.

ROMNEY—National Guardsmen clean debris from this Hampshire County bridge. (Courtesy of Hampshire Review)

ROMNEY—The riverfront Blue Beach Camp was almost completely destroyed by the South Branch. (Courtesy of Hampshire Review)

ROMNEY—Water from the South Branch undercut parts of many roads in Hampshire County. (Courtesy of Hampshire Review)

ROMNEY—Pendleton County residents inspect the remains of this home. (Courtesy of Hampshire Review)

Panic day: 'Two in each boat . . . one to row, one to scream'

Nov. 20
The Weston Democrat
Weston

Panic: a sudden over-powering fright; esp: a sudden unreasoning terror often accompanied by mass flight.—Webster's New Collegiate Dictionary.

It was a few minutes after 4 p.m. Tuesday afternoon, November 5, near the end of the second and worst day of a history-making flood that was ravaging much of northern and eastern West Virginia, including Weston and Lewis County. Bill Woodburn, Resident Construction Engineer at the Stonewall Jackson Lake Dam, three miles up the West Fork River from Weston, was sitting behind his desk in his office at the U.S. Army Corps of Engineers field headquarters, a few hundred yards down-stream from the huge, concrete barrier, hoping for a few minutes of rest. He'd been awake and at work for almost 36 hours, "nursing his baby" through its first crisis and limited test.

Suddenly, the office door burst open, and Lou Hutchinson, Project Manager for the dam contractor, rushed in with some disturbing information. He had just spotted several pieces of structural timbers floating on the floodwaters below the dam. Woodburn, believing he knew where the boards had come from and therefore, fearful of the implication of Hutchinson's discovery, left at once in his vehicle to inspect the dam and see for himself what was happening. He had been in the area of the dam's base only a short time before, and all appeared well.

The river had crested both above and below the dam about 1 p.m. that afternoon and had been falling slowly since then. At 3 p.m., the engineer's flood gauge on the downstream side had read 1027.15 feet. Now, at ten minutes after 4:00, the water level at that spot was back up to 1028.20, a rise of just over one foot in an hour's time. Something had happened, that was certain, and probably in only the last 10 or 15 minutes, Woodburn thought, to allow more water out of the dam, still under construction and not yet able to hold back more than a small fraction of the runoff from Hurricane Juan's heavy rains, which had fallen since Sunday morning, a frog-strangling 7.15-inch deluge, an all-time record for a 48 hour period in the immediate Weston area.

The engineer concluded at once that his initial fears were facts; there was no other logical explanation for the pieces of timber he saw rising and dipping on the muddy waves. A downstream-side wooden structure used to form up the concrete for an electric power generation penstock at the base of the dam, on its east end, had washed out. And a ribbed steel plate temporarily capping the upper end of the 60-foot-long tunnel, deep under the coffee-brown waters and therefore not observable, had likely ruptured, allowing substantially more flood water, water that earlier had been held behind the dam, to flow through 13-foot-diameter opening. Just how much more was now gushing into the torrent below the dam and heading for Bendale and Homewood he could not estimate, having at hand no measuring instruments or interpretive charts for that purpose, but it was enough, he thought, to cause the river to begin rising again at least as far as Weston.

Woodburn bounded back into his truck and sped for his office —and the telephone, wondering just what to say to alert authorities downstream to the new danger. The West Fork had crested in Weston before dawn that day at 24 feet, three inches. By 4 p.m., when the penstock problem arose at the dam, the river had fallen 15 inches, to 23 feet. The emergency appearing to be over to most of Weston's telephone exchange subscribers, the volume of their calls had now subsided, and it was now possible to get a dial tone almost instantly, compared to the one-to-three-minute delay on Monday and Monday night. The engineer placed his first call to the Weston Volunteer Fire Department, where it was received by Assistant Chief Ed Griffin at 4:15 p.m. Chief Griffin said afterwards he thought the engineer, to whom he had spoken on the telephone numerous times during the preceding 24 hours, sounded unusually excited and uneasy this time, and, later, Woodburn concurred that he was indeed apprehensive and, no doubt, his feelings were reflected in his voice. Both agree that what follows is an accurate narration of their conversation.

Woodburn to Griffin: "We have had some trouble here at the dam, a failure of a retaining form in the penstock, and the water has raised here about a foot in the last 15 or 20 minutes. I don't know what it will do to you, but I expect you'll get more water in Weston. We'll monitor it every five minutes and keep you informed."

Griffin to Woodburn: "How much rise do you think we'll get?"

Woodburn: "I don't know"

Griffin: "Will we get a foot more?"

Woodburn: "I think not more than a foot or two, but I really don't know."

Griffin: "Could we get another ten feet?"

Woodburn: "I really don't know. The water is rising here; that's all I do know right now. We'll be monitoring the situation, and I'll call you back every five minutes."

He hung up. Woodburn and Chief Griffin are in accord that the engineer never said the dam itself had burst, because, of course, it hadn't. Nor did he say that a "wall of water" was coming downriver, or that Westonians had just so many minutes to get to high ground. Nor did the Chief make such statements to anyone at any time. Those false and panic-creating reports were left for others —not well-informed in most cases, and perhaps ill-intentioned in one or two instances—to make and spread a little later.

Because the engineer was less than certain about the effects of the penstock water release on Weston, Chief Griffin decided he had no alternative but "to pull the plug," as he later put it: to order the city of Weston evacuated. "Given the same circumstances," he said, "if they were to arise again, I would do the same thing in the same way. I didn't want the responsibility for hundreds of deaths."

Turning from the telephone—red-faced, one observer noted—he told about a dozen firemen who were milling about in the lounge next to the department's communications center, that "some kind of internal thing at the dam has broken. We're going to get more water, and they don't know how much." In a later "post mortem" of the dramatic event, Chief Griffin said it was a mistake for him to use "dam" and "broken" in the same sentence, as probably those were the only words some of his listeners heard. But he didn't have time to think of that at the time. Nor did he name the dam in trouble, assuming everyone would know it was Stonewall Jackson. And not everyone did, it turned out.

With no delay, he began barking out orders, dispatching men to various points throughout the community. Herbert Butcher and Gary Bennett were sent to the West Virginia Glass Specialty Company in Homewood with a two-way radio and a yardstick to measure the rise of the water in the highway there. He cautioned them to take a safe position from which they could run to higher ground if the water should raise dangerously fast. Jonathan Alfred, with a walkie-talkie, was sent to the old fire department building on Water Avenue, where he could observe the river gauge on the wall of that building and report back to the station. In floods in earlier years, the gauge had been televised over a cable system channel.

The remainder of the men were told to begin going door to door in the main part of town and in west Weston, or to use public address speakers on their cars and trucks, if they had them, to warn people to evacuate. The Chief then reported the crisis on the fire department's two-way radio system to alert his men who were working at flood-related tasks in various parts of town. He knew that message would get through to and inform hundreds of other people with scanner radio receivers. He instructed those firemen in or with boats to move them to protected spots and to stand by to use them in rescue work, if necessary. Next, he made an announcement on the department's FM radio channel on the TV cable system; it would be heard in firemen's homes and warn their families. Then he punched in the buttons that turn on the fire sirens at the water company's filtration plant and at the City Building.

There were four men left at the fire deaprtment: Chief John Shea, Assistant Chief Griffin, Director of Emergency Services Bill Rowan, and Fireman Kevin Kiddy. And two boats. "If we had to evacuate the station, two would go in each boat." Chief Griffin later explained, "one to row and one to scream."

The Chief does not remember now whether he phoned the radio station, or if someone at WHAW called him. Nor is anyone at the station sure. It may have been the latter, because the station monitors the FM channel and runs a scanner as a part of its local news operation. Whichever way it was, he talked to Station Manager Arkie Posey, told him what had happened, and asked him to begin broadcasting a warning to Westonians to move quickly to higher ground. The station's first message was put on the air around 4:20 p.m. by J.B. Parker, recalls Don Detamore, who was in the studio, also. The substance of it was that a large surge of water was coming down the river from the Stonewall Jackson Dam, and people should evacuate low-lying areas immediately. WHAW-WSSN Radio repeated that warning three or four times in the next 15 or 20 minutes, interspersing the announcement with music and commercials.

Meanwhile, volunteer firemen were racing into those areas not blocked by the still-flooded river to give warning. The City Police Department, the Lewis County Sheriff's Department, and the Lewis County Emergency Squad, all of whom had heard the early reports on their scanners, were moving to take similar action.

A "Democrat" reporter was one of the first persons on the street to be informed of the emergency. Standing on the Bennett corner of Center Avenue and Bank Street, he was talking to Hunter Bennett about the flood when Freddie Potter, a member of the fire department, drove up and said: "Some part of the dam has broken loose, and we're expecting a lot more water in the next few minutes. Everyone is being urged to evacuate to high ground." The reporter, who had been hearing false reports of that nature since the flood began, said, "Baloney!", or, more accurately, he used a two-word expletive never printed in a family newspaper. "No, this is the truth," Potter emphasized. "We just got a call at the station from the dam, and it's been verified."

The reporter believes, but isn't sure, he said something in the way of "farewell" to Bennett and then walked down Bank Street to Main Avenue, where he saw the first signs of panic. Possibly seven or eight people were running—up the street, down the street, and across the street—in both directions.

A couple of trucks came tearing around the corner of Second Street and roared up Main toward First at breakneck speed. At that point, he began running himself toward the newspaper office, where Editor Bob Earle was working alone on that week's issue—most of the other employees were flood-bound—and was probably not aware of the new situation. In front of Murphy's, he was slowed down by two ashen-faced ladies who asked him where they should go to be safe. "Go up to the Columbia Club. That ought to be high enough," he told them, not pausing to ask if they knew where it was.

When the reporter reached the newspaper office, he found the editor on the telephone, talking to a man in Congressman Bob Wise's office in Washington. The Congressman wanted to get some information in the paper's next issue concerning applications Lewis County's flood victims could make for grants and low-interest loans to mitigate their property losses. After a minute or two, the reporter decided he had better interrupt the conversation and tell the editor what was happening at the dam, and to suggest both of them ought to get to higher ground. The Washington caller heard that on his end, too, and told the editor to call him back when he could.

The newsmen then left the "Democrat" office, planning to drive their vehicles separately to Earle's high and dry home on Town Run Road. On the way, the reporter stopped long enough at Murphy's,

WESTON—Five feet of water inundated Tucci's Store in Lewis County. (Courtesy of The Weston Democrat)

WESTON—Second Street in Weston is covered by water. (Courtesy of The Weston Democrat)

where they were trying desperately to get the store closed, to buy several rolls of film. "If we were about to have a catastrophe," he said later, "I wanted to get pictures of it." One of the clerks in the store who was running for the door told him, "Go ahead and get your film! You can pay us later!"

Following are the personal experiences and reactions of a number of other Lewis Countains who were caught up in what has to have been the most dramatic and exciting hour in the community's history:

POLICE CHIEF GEORGE BLAKE—"I'd just got to my home on Cottage (Avenue) from the station, where I'd been since the flood started the day before, and I got a phone call from the desk sergeant. He said the fire department alarm radio had just reported that the dam had broke or malfunctioned, and the town had to be evacuated. I ran out of the house with my wife. We jumped in my truck and headed for the City Building. I stopped at the IGA to tell them what had happened and sent my wife on ahead. 'Tell everyone you see,' I told her. When she got to the station a few minutes later, she went upstairs to the mayor's office to answer his phone. When I got to the station, I immediately called his house to tell him what was happening.

"Then I had some of our men drive the city cruisers and trucks up on Kent Kessler's hill while I helped answer the phones. When I could, I went outside and watched for a two-foot-high wall of water I thought was coming. Fire Chief John Shea called and told me we should evacuate the building, but I told him we would hold the fort down as long as we could. Then some girl came over from the Consolidation Gas offices and said one of their men was up at the dam and there was no cause for alarm. The river there was up maybe six inches, that was all. About that time, I got a call from someone who said they were phoning from the President's office in Washington. They wanted to know if it was true the dam had broken. CBS and NBC called to ask the same thing, and newspapers from all over the country. I told them all that I thought the dam had malfunctioned, not broken, but at the time I could neither confirm nor deny anything."

WESTON—Stop signs were covered along Fourth Street and Center Avenue by water from the West Fork River. (Courtesy of The Weston Democrat)

MAYOR DANNIE WHELAN—"I had just gotten home and was in my taxidermy shop, getting ready to preserve some animals that had been frozen but were now thawing out. I phoned one of my buddies who lives in the Kitson Apartments on Main Street and while we were talking, I heard the fire whistle begin blowing. Then over the phone, I heard a fire truck with its siren going by my friend's apartment. He told me it was heading my way. After a minute or so, when it hadn't come up Pike Street, I asked him to go outside and see if there was something going on near his place.

"When he came back to the phone, he said he had just seen George Blake running by, waving his arms, and yelling, 'My God, the dam has broke! We've got to get people out!' I panicked. I thought, 'My God, I'll lose my shop.' But then I calmed down and started running door to door in Kitsonville, screaming for people to get out and onto higher ground. I had this awful feeling of helplessness. There were just too many houses and too many people to get to them all. But by the time I got to Kitson Street, I realized most everyone had already gone up on Jackson and were probably safe. Then I went on to the City Building and saw that George had things under control there. So I cruised the city to lend a hand where I could. I'll tell you what, it was an experience. But I'm proud of the way we handled it. It was pretty well done."

DEPUTY SHERIFF MARGARET ADAMS—"I was in the jail office with Lou Ann Mick, our correctional officer, as we had been since Monday, trying to keep things going there while our deputies were out on the Interstate rerouting traffic around the flood water at Milepost 105. I thought I heard something on the scanner, something about the fire department telling all units to go to high ground immediately. I asked Lou Ann if she heard it, and she said, 'No.' Then they repeated it. I went over to the back of the courthouse, to the Commissioner's room, where some of the ambulance people had set up an emergency center. I asked Martha Williams if they had heard something about evacuating, but they hadn't. So Janice Peters called the fire department. Ed Griffin answered, and all he said was, 'Get to higher ground immediately.'

"So I went back to the jail to call Sheriff Hall on the radio. He had just left a little bit before to drive up to Jane Lew to see how things were going. I told him we had been ordered to evacuate, but I still didn't know why. He said to get the prisoners out of their cells and take them up on the hill, to High Street or somewhere that was safe. We were in the process of doing that when Martha Williams ran in and reported the dam was giving way. Then I got on the air and told all the deputies to get to higher ground. It was awful. The Johnstown flood catastrophe came to my mind. I could just see elderly people and children tumbling and tumbling over and over in flood water. I was really scared for what might happen.

"Deputy Dave Ireland, who had been going since early Monday without any rest, was asleep with his dog in a cruiser behind the jail. We woke him up to help Lou Ann and me, Paul Fleming, and our cook, Charley Marshall, get the prisoners out and up the hill. There were ten of them in jail at the time. We had them all lined up in their orange uniforms along the wall above the library parking lot. After a while, Sgt. Mike Matthews and Deputy Robert Snyder came in from Interstate patrol to help us guard the prisoners, and then Claude Riffle gave word that there was just going to be a slight raise in the river. So we brought the prisoners back to the jail and fed them their evening meal."

SHERIFF M. RALPH HALL—"I was on the Interstate, coming north from the Roanoke exit, when Margaret Adams called me on the radio about the evacuation order. Then I heard on WHAW that something had happened at the dam and a lot more water was expected to come down the river.

I thought the best place I could be was the bridge over the river at Exit 96, just below the dam, where I could park and watch the river and report any wave of water or anything else that might happen. Later, when the radio station said there was no longer any danger, I came on into town to look after other problems. I'm so proud of all the work my deputies and jailers did during this whole

WESTON—Water rose into the stately home of Harry Underwood on Center Avenue. (Courtesy of The Weston Democrat)

thing. They're a great bunch of men and women; they really put out for the people."

JAMES GLADKOSKY, County Commissioner Coordinator—"I was in my apartment on the second floor of the Keener Apartments on Third Street, and a friend called me on the telephone. He said he'd just heard on the police radio channel that the Stonecoal Dam had burst and there was going to be a rush of water into Weston. So I called the sheriff's office, and they told me it was a break in the Stonewall Jackson Dam; there'd be a big influx of water, and I should evacuate. I called the fire department for a boat, but Bill Rowan told me there were no boats then. 'You're on the second floor,' he said, 'that ought to be safe. Don't worry. There's nothing to worry about.'

"So I kept going downstairs to check the water, measuring it, and I couldn't see any rise at all. Later on, I heard someone on a bullhorn: 'Evacuate now! Evacuate now! Get to higher ground.' I looked out and saw Dr. Windon on down the street at his office. There was a boat there, and they were telling him to hurry, hurry, get in. It was five o'clock then, and, as I learned later, it was all over by that time, although not many yet knew it. Later, I heard some criticism of Ed Griffin, but he used his best judgement, based on what he knew, and as far as I'm concerned, he did exactly the right thing."

KIM OLDAKER, F & M Lunch Employee—"We had a restaurant full of people, and all of a sudden this fireman came running in and hollering, 'The dam broke! Everybody get out! Get to higher ground! There's going to be 20 feet of water coming.!'

"Well, Sherri Giles and I told everybody to just leave. Don't worry about paying. As fast as we could, we locked up. We left everything on, the stoves, the grill, the lights, and everything. We joined a lot of people who were running up Bank Street, yelling and screaming, 'What are we going to do?' They were stumbling and knocking each other down. It was just crazy."

GRACE WEINBERGER—"I was at the post office corner of Bank Street, and all of a sudden all hell broke loose. People—at least 30 of them—were running up the street from Main like a herd of cattle. I never saw such panic in my life. I'm the calm one, so I just stood and watched them. They were yelling at each other, 'The dam broke! The dam broke!' I asked, 'Which dam?' They didn't stop to answer. Then Freddie Potter told me it was something at the Stonewall Jackson Dam that had broken, not the dam itself, and I wanted to go back to the Christmas Town store, but he didn't think it was wise just yet.

"Then I met this woman running up the street. She was out of control, crying, sobbing, her chest was heaving. I thought she was having a heart attack. I put my arm around her and tried to comfort her, and I talked her into going with me to the Columbia Club, where maybe we could help others who were in more trouble than we were. Once we got there and things were not so frantic, she said it was the best thing to happen to her; I mean, going up there to help out. You know, it was nice in a way. People who did not know each other were talking to one another, consoling and comforting them just like family. Then John Stalnaker came into the Club and said we should get out of there, that we should get up on Bennett Avenue. When we got up there, Gil Miner and I stood talking to each other and just watched others as to how they reacted."

Meanwhile, back at the dam, Engineer Woodburn had called his superiors in Pittsburgh, Pa., and Huntington to tell them what had happened. They put a hydrologist on the line to take down what little data Woodburn had, hoping to give him a forecast of what new crest to expect on the river.

Post Script: When all the alarms and excursions were over, and concrete was being poured again to fill the remaining gap in the dam, Engineer Woodburn was asked how safe the Stonewall Jackson Dam would be when completed. "Safe?! Safe?" he responded. "Hell, flood water could pour over that thing for 25 years, and I wouldn't hesitate a second to sleep on it!"

WESTON—Ronnie Radcliff surveys the floodwaters that surround his house. (Associated Press Wire Photo Courtesy of Huntington Herald Dispatch)

MORGANTOWN—Waist deep water faced the residents of Short Street in Granville. (Courtesy of Dominion Post)

Friends help Marlinton hospital 'recover'

Nov. 17
Morgantown
By J. Richard Toren
University News Service

Pocahontas Memorial Hospital hopes to be back in the business of caring for the sick and injured by the end of the week, thanks to an outpouring of help from some special friends.

Just 12 days ago Administrator Larry Loftin told a worried telephone caller, "My hospital is a river bed at the moment." At that point the angry Greenbrier River had submerged the entire ground floor of the hospital half a block away.

But as this past week neared its end, a wearied Loftin could speak hopefully. "We hope to be back in service on at least a limited basis by the end of the next week and admit selected patients."

Much of the service will be provided from new modular units to be installed adjacent to the hospital, including X-rays, emergency and laboratory service, and probably food for patients and employees.

All of those units had been located in the ground floor and were effectively wiped out by the swift, muddy flood waters that submerged it the night of Nov. 4 and also climbed another foot into the first floor.

Recovery efforts have been spearheaded by skilled technicians from the West Virginia University Medical Center and WVU Hospital in Morgantown who restored heat and full electric power this past Wednesday. They are still on the job or on call as work proceeds to get the modular units delivered, in place, and equipped for patient use.

Fourteen WVU electricians, millwrights, carpenters, plumbers and heating ventilation workers went with the first wave of expert workmen Tuesday, after a team of four administrators made the eight-hour round trip Monday to assess damage and determine what was needed.

Pocahontas Memorial, one of West Virginia's smaller hospitals with 40 beds, is a three-story red brick structure located downtown and practically on the banks of the Greenbrier River.

To say it had "grief and trouble" understates the case. It is an old building to begin with, the main structure perhaps 65 years old. On the end near the river is a patient-care section added in 1950. A World War II field artillery piece stands out on the front lawn.

Loftin had only hours notice last Monday that the flood waters would reach and threaten the hospital. By 8:30 p.m. they were in the ground floor and the power had to be shut down to prevent the transformer boxes from exploding. More than a week later most clocks in the building still showed the ominous 8:40 time.

"The current was terribly swift, and running on one side as well as the front lawn and the back parking lot," Loftin recalled. "We moved patients to the upper floors and they weren't in danger, but there was no light but flashlights. And lights were going off over town. We could see some of the explosions as the boxes blew up.

"A friend got through by telephone to ask how things were going, and I told him, 'My hospital is a river bed at the moment.'

Engulfed in the torrent were the electric utility room, elevator shaft pit, laboratories, X-ray facilities, emergency room, kitchen and work and storage areas—almost everything needed to run the hospital.

The WVU Medical Center got involved in the restoration project when John E. Jones, vice president for health sciences, contacted the State Health Department asking where Medical Center help was most needed. Within hours the administrative team was on its way to Marlinton and the commitment firmed up.

Most of the muck and mud had been removed by a commercial

MORGANTOWN—Mike and Rhonda Hamilton remove microwave oven from their flood damaged home on Van Voorhis Avenue in Morgantown. (Courtesy of Dominion Post)

firm and partial electric power was restored by an area contractor before the WVU team arrived.

"They had done a terrific job of cleaning up before we arrived," said Franklin Wolf, leader of the crew of technicians. "Our first job was to get the boiler going so we could restore heat and let the building start to dry out.

"We did that by Tuesday midnight, but the transformers failed after one cycle and we had to get parts from Lewisburg the next morning."

Millwright Grove "Buck" Pennington noted that, "It's always easier to put things together if you've taken them apart yourself. In this case, some parts from the boiler were piled together, and others had been torn loose and moved around by the current."

Some of the most exacting work fell to the electricians. The large transformers had to be moved, cleaned and baked to remove moisture—work done by local contractors—then rewired by WVU craftsmen into a jungle of smaller wires serving the whole building.

"We've been using industrial-strength 'hair-dryers' to dry individual switches and transformers," said chief electrician Dana Sypolt.

By Tuesday evening Blehschmidt and Harold H. Harper, Dr. Jones' special assistant, had been to Pittsburgh to look for modular trailers. Both had been involved in Monday's assessment visit along with physical plant chief Gary Miller and George Mikitarian, assistant administrator of WVU Hospitals, Inc.

"We found a unit that could be adapted quite well for a temporary emergency room, with double doors opening at just the right location for emergency vehicles," Harper said.

One modular unit was expected to be delivered this weekend and a second this coming week.

Mikitarian has been working with Loftin on administrative matters, including creation of a task force to consider the future of the county-owned hospital.

Loftin said he is hopeful that the new bonds being forged with the WVU Medical Center will be continued when the present emergency ends.

MORGANTOWN—National Guardsmen and volunteer firemen work to sandbag the Morgantown Sewage Treatment Plant. (Courtesy of Dominion Post)

MORGANTOWN—(Top) Debris backs up behind Lake Lynn Dam of Cheat Lake near the West Virginia/Pennsylvania border. Workers in the foreground are collecting the barrels to inspect for toxic waste materials. (Bottom) The Lake Lynn Dam pours millions of gallons of water through its spillway as the refuse backs up behind the Cheat Lake hydroelectric works. (Both photos courtesy of Dominion Post)

ELKINS—Flood refugees paddle along what once used to be a roadway in Crystal Springs. (Courtesy of Dominion Post)

ELKINS—National Guardsmen and volunteers form a chain gang and remove damaged items from homes on Ellis Avenue Nov. 6. (Courtesy of Dominion Post)

ELKINS—Mobile homes in Bowden were pushed aside by the swift waters. (Courtesy of the Inter-Mountain)

ELKINS—Verle E. Harper examines his cabin in Bowden, which has now become a perch for a mobile home. (Courtesy of Dominion Post)

GLENVILLE—There was no traffic Nov. 5 in Glenville when the light (center) turned green on one of the main streets. (Courtesy of Pathfinder)

GLENVILLE—Like hundreds of other residents, this unidentified man faced the hard chore of cleaning up a flood-soaked town. (Courtesy of the Parkersburg News)

GLENVILLE—The Go-Mart in the downtown section received heavy flood damage which forced the disposal of much of the store's inventory. (Courtesy of Herald Dispatch)

GLENVILLE—The Kanawha Union Bank was one of many businesses invaded by the Little Kanawha River. Later, workers began the task of cleaning out damaged furniture and records. (Courtesy of Pathfinder)

GLENVILLE—This bridge over the Little Kanawha River withstood a partial submersion and later served traffic. (Courtesy of Pathfinder)

What goes on in a federal disaster center?

Nov. 21
The Morgan Messenger
Berkeley Springs
By John Douglas

A federal disaster center is now open in Martinsburg to serve flood victims in Morgan, Berkeley, Hampshire and Jefferson counties. The center at the armory at Martinsburg airport will be open from 9:00 to 6:00, Monday through Satuday, until further notice.

Last Friday I got a taste of exactly what the flood victims will be going through. With a Morgan Countian whose home and farm were damaged, I went to the disaster center then set up at the School For The Deaf & Blind in Romney.

After you make your way through a federal disaster center, you're likely to be pretty confused. You'll have spent a few hours filling out forms with the help of government employees. When you leave, you're liable to have still more forms in your hand. You may wonder if it's all worth all the hassle to do all that paperwork when you could be cleaning up your house.

The answer is: **Yes, all the inkwork will probably be well worth it for most people.** No, the government is not out to further flood you in a sea of red tape.

Meet FEMA

After a receptionist takes your name, the first person you'll meet at a federal disaster center is the FEMA Man. He works for the Federal Emergency Management Agency, otherwise known as FEMA.

FEMA is the federal agency charged with dealing with crises such as major floods. East Coast offices are presently so overworked that the employees in West Virginia disaster centers are from California, Colorado and western offices. The signs around the room are in both English and Spanish.

The FEMA Man has faced such situations before. He'll talk you through a couple applications, which he'll even fill out. He'll get the story about precisely what you lost in the flood.

He does the applications for emergency housing if you can't stay in your flood-damaged home. The housing applications are sent on to Clarksburg each day. It should take about a week to get word back.

Flood Insurance

Clarksburg is the clearinghouse for aid applications from all over the state. It's also where federal officials are deciding on flood insurance claims for those who were insured. The vast majority of Morgan Countians, and West Virginians, weren't.

The FEMA Man requires all aid applicants to sign a statement regarding flood insurance before he'll talk about low interest government loans. Essentially, you have to agree to have flood insurance if you use federal loan money to rebuild in the floodplain. Annual flood insurance premiums are currently $4.50 per $1,000 for buildings and $5.50 per $1,000 for contents.

Federal and state officials would prefer to steer people away from putting their homes right back in the floodplain. New structures are subject to floodplain building restrictions.

Talking Money

The FEMA Man is now ready to talk money. Outright grants of $5,000 are available to low income families. Small Business Administration loans of up to $100,000 are available at 8 percent interest to those with established credit. Loans of 4 percent may be available to some who couldn't otherwise get a loan.

There are those who won't qualify for loans. In such cases, the $5,000 grant may not replace what they lost. Helping such people is currently a political question that Senator Jay Rockefeller and Congressman Harley Staggers Jr. are working on.

The FEMA Man will not fill out the loan application for you. The somewhat complicated form must be taken home and filed no later than early January. **There appears to be a real need for responsible people who can aid West Virginians in filling out the loan applications.**

Grants and loans are not available for those who lost second homes or vacation property.

Before he sends you on, the FEMA Man will list a number of other people you should see while you're in the disaster center.

You may be sent to a mapreader who will determine where your property is located in the floodplain, or if it's in the floodplain at all. Some flooded areas were never hit before. The flood in Pocahontas County, for instance, is considered a 500 year one.

IRS Can Help

Next you'll talk to the Internal Revenue Service, and this time the IRS can truly help. Flood losses can be tax deductions over the course of years. You can even get a refund back from the taxes you paid last April.

The IRS Man won't fill the form out for you but he'll give you a workbook with a memory-jogging method of calculating your flood losses or, in this case, your tax deductions.

When calculating losses, a qualified appraiser can be a big help.

The tax side of things is especially important. If you apply for a low interest loan, you must also provide the last two years worth of income tax returns.

Steve Fielder of Martinsburg is organizing a group of accountants, attorneys and others to help with flood victim's questions. Fielder's an associate with the law firm of Martin & Siebert.

If you lost deeds, car titles or other legal documents, you'll be sent to a Legal Services representative who has forms at the disaster center for document replacement.

The West Virginia State Bar Association is also running a telephone hotline to help handle legal questions.

The Red Cross, ever present in West Virginia these days, is at the disaster center too. They'll ask about your immediate needs. At Romney, people were given brooms and jugs of Purex.

Other agencies at the centers are more specialized, depending on particular needs. You'll find representatives of the Department of Human Services, agriculture agencies, Small Business Administration, and others.

Worth It

If you suffered losses in the Flood of '85, the trip to a disaster center won't exactly be a fun afternoon. Still, there is the real chance you'll find some help.

It may be excruciating to walk out with a sheaf of government forms in your hands, but that's the way the government works—a form for everything, and **nothing without a form.**

PAW PAW—A view of Paw Paw and the flooded Potomac River on Nov. 6. (Courtesy of The Morgan Messenger)

PAW PAW—Nothing but debris remains in Consolidated Orchards. (Courtesy of The Morgan Messenger)

PAW PAW—The rain swollen Potomac River rose to engulf Paw Paw. (Courtesy of The Morgan Messenger)

PAW PAW—Many railroads in the Potomac Valley and throughout the state were closed because of high water. (Courtesy of The Morgan Messenger)

"We were in the tree for seven hours."

—— Six-year-old Adam Snider, flood survivor

PAW PAW—Two towers, which once held up the center portion of Paw Paw's temporary bridge were swept away by the Potomac River. (Courtesy of The Morgan Messenger)

PAW PAW—Isolated because of the bridge washout over the Cacapon River, Sue Jordan, Scott Jordan and Dustin Gardner use a makeshift raft to travel back and forth across the river. (Courtesy of The Morgan Messenger)

PAW PAW—A roof from a cabin in a camp area on Cherry Run smashed against another dwelling. (Courtesy of The Morgan Messenger)

PAW PAW—The Great Cacapon River overflowed into the Green Acres Campgrounds along Rt. 9. (Courtesy of The Morgan Messenger)

Record levels reported at dam

Nov. 9
Mountain Statesman
Grafton
By Bob Bucy

A new record was set at the Tygart Lake Dam on Thursday, as the pool crested at an all-time high.

At 8 a.m. on Thursday, the pool stood at 1156.67 feet above mean sea level. The previous all-time high was recorded in June of 1972, when the pool stood at 1155.00 after Hurricane Agnes went through.

PENCE SPRINGS—This dwelling was among 152 which were destroyed in Summers County during the November 1985 flooding. (Courtesy Huntington Herald Dispatch)

Grafton coming to aid of others

Nov. 7
Mountain Statesman
Grafton
By Bob Bucy

Grafton is coming to the aid of its neighbors.

The city is taking contributions of non-perishable food items like canned goods, cleaning supplies, blankets and cots and other items to provide aid to flood victims in surrounding counties. Donations are also being accepted, which will go towards the purchase of emergency relief items. The city is working in conjunction with the Salvation Army, which is delivering the goods to the surrounding counties.

"Historically, Philippi has been good to the city of Grafton and they provided us support in the water shortage a few weeks ago when other cities cut us off. We have an obligation to help those people in their time of need," Mayor Paul Elder said. "We're also doing all we can to help aid Preston Countians as well," he said. Elder said the city has received a "good response" thus far in collecting the non-perishable items, as area residents, businesses and churches began making contributions at City Hall on Wednesday. "We shipped the first truckload to Philippi on Wednesday and that's a big step. Now, we need to identify what exactly the biggest needs are and make contributions in those specific areas."

Elder said blankets are a key item on the list. "I've talked to several state officials and they tell me that blankets are a major need and we're going to see what we can do about getting some," Elder said. The mayor contacted a blanket firm in Cumberland, Md., which is involved with a printing firm in Parsons and also Georgia Pacific. "They're going to donate 100 blankets and they're also working up a price for us on the cost per 100," Elder explained. "Hopefully we can begin getting donations from around the area to help in the purchase of these. We have a moral obligation to help these people and hopefully we can raise this money," he said. "The victims will begin to feel the real brunt of this in the next few days when they try to start their homelife again. They'll be under some real pressure and we've got to get geared up now to help them."

Lt. Charles Smith of the Salvation Army was busy Wednesday loading up vans and other vehicles to distribute goods to the victims. "We're involved with distributing the items for the city, along with coordinating relief efforts in Barbour, Preston, Randolph and Tucker Counties," Smith said. He said that a Salvation Army Canteen left Wednesday afternoon for Elkins, while two vans were headed for Philippi and Belington. The canteen is a trailer which can store goods and hot meals can also be cooked inside.

PENCE SPRINGS—Only the pillars remain from the span which crossed the Greenbrier River here. The flood washed the bridge nearly 100 yards downstream. (Courtesy of Huntington Herald Dispatch)

HINTON—This destroyed summer camp came to rest in a grove of trees after the Greenbrier River pushed it off its foundation and down the river on Nov. 6. (Photo by Beckley Register Herald, Associated Press Wire Photo courtesy of Huntington Herald Dispatch)

Wedding present

Nov. 8
Evening Journal
Martinsburg
By Teri Wells

Todd and Debbie Shaffer had not even settled into their new home before the churning waters of the Potomac forced them to abandon it.

The Shaffers, who were married Oct. 26, had not unpacked some of their wedding gifts. On Thursday, the once gleaming appliances and kitchen utensils were lifted from the muddy kitchen shelves and loaded into the back of a pickup truck.

"Isn't this a wonderful wedding present?" said John Shaffer, Todd's father, as he pushed a small wall of slippery mud out the door.

A portion of the river bed had found a new home inside the four room house. Satin bows and wrapping paper from the wedding presents, once crisp and colorful but now a uniform dull brown sheen, peppered the living room floor.

At the flood's peak, the swirling waters had risen to the house's roof. The house, in the Falling Waters Mobile Home Park, stood on firm but soggy ground when the floodwaters receded.

A trailer next door, owned by John Shaffer, had been ripped from its supports and deposited about 20 feet from its original site.

"We were able to get our clothes out before the flood hit," Debbie Shaffer said as she placed an unused crock pot into a wheelbarrow.

Todd, 21, and Debbie, 20, stacked their dripping furniture beside a tree near the porch of the house. The Shaffers, with the help of family and friends, began the massive cleanup effort of their rented home Thursday morning.

"We're going to be at this awhile," Todd said.

After the truck was filled, the muddy items were taken to John Shaffer's residence for a thorough dousing with a garden hose.

"The only way you can survive this is to keep your spirits up," John Shaffer said.

Despite the grim nature of the exhausting task, the mood elsewhere in the trailer park was not totally overcast. Energetic voices rose above the still angry roar of the Potomac as residents and the cleanup crews, recruited from among friends and relatives, began to pick up the pieces.

"I found the doorbell . . . now if I could just find the door I'd be okay!" one man said.

"How you doing, neighbor . . . we've got a bit of a flood here," another called.

Jesse Schissler found that the pier in front of his trailer had washed away, but his boat remained securely tied to a tree.

"I can't believe it didn't wash away," he said.

The river, usually about 50 feet away from the trailer, was now only a few yards away.

Schissler and Charlotte Byrd viewed his vacation home Thursday morning for the first time since the flood hit. The two opened the trailer door to find a virtual sea of mud and debris within. Small fish flopped helplessly in the sticky mixture.

"I always liked to make mud pies when I was a kid, but this is a little ridiculous," Byrd said, stepping carefully inside and surveying the damages.

Schissler removed sections of insulation that had dislodged from the roof of the trailer and soaked up the mud like sponges. The wooden-paneled walls of the trailer, as well as all furnishings and items in the interior, were coated with the thick mud.

With little other talk, the two returned to the same grimy job that residents all along the river were experiencing.

LITTLE GEORGETOWN—These Berkeley County women clean up their mud-soaked kitchen after floodwaters receded from West Virginia's Eastern Panhandle. (Courtesy of The Evening Journal)

Rivers of destruction loosen their grip

Nov. 7
Martinsburg Evening Journal
Martinsburg
From staff reports

Massive floodwaters began receding in the tri-county area Thursday, leaving behind homeless residents, millions of dollars' worth of damage and widespread destruction.

Weather officials say the flood is worse than that spawned by Hurricane Agnes in 1972, and may be recorded locally as the second or third worst this century.

Paw Paw appeared to be the hardest-hit community, where floodwaters swamped a 25-lot mobile home park and destroyed many businesses and homes. The temporary bridge linking the town with Maryland was heavily damaged.

The Harpers Ferry National Historical Park has suffered roughly $250,000 in damages, according to park superintendent Donald Campbell. A number of homes in the Bloomery and Millville areas were submerged as residents watched from the banks.

Berkeley County Commission President Vic Shockey declared a state of emergency as many residents fled their homes in Little Georgetown and surrounding areas.

Flood levels surpassed those created by Hurricane Agnes and rivaled a 1942 flood where the water crested at 32.7 feet in Harpers Ferry. The flood stage is 15 feet in Shepherdstown and 18 feet in Harpers Ferry.

The 1935 flood pushed water levels to 42.1 feet in Shepherdstown and 36.5 feet in Harpers Ferry.

Floodwaters crested at 25.7 feet in Millville and 30.6 feet in Shepherdstown at 1 a.m. today, according to Laurence Lloyd Sr., local weather observer based in Charles Town.

National Weather Service officials said Harpers Ferry had a flood crest of between 33 to 34 feet—more than 15 feet above flood stage.

Flood-related incidents and actions have been numerous in the tri-county area in the past two days:

—Many Berkeley County residents, some armed with shotguns, remained on their water-ravaged property to protect their belongings from looters.

—A 34-foot-long sandbag wall constructed at the bottom of High Street in Harpers Ferry protected some adjacent shops from water and debris damage before it partially collapsed Wednesday afternoon.

—Commuter train service was shut down for the second consecutive day because of floodwaters that threatened bridges and inundated tracks.

—Paw Paw officials were trying to keep more than 100 evacuees from returning to their homes Wednesday so crews could stabilize leaking propane gas from two 18,000-pound tanks that broke loose from Vesuvius Cruciable.

—Agriculture officials said between 100 and 150 acres of corn were destroyed in Morgan County.

—One Jefferson County mobile home was uprooted from its foundation and floated down the Shenandoah River before striking several trees.

TAYLOR COUNTY—A record fill was chalked up at the Tygart Dam as water creeped further into the parking area. (Courtesy of Mountain Statesman)

TAYLOR COUNTY—The flood of '85 is believed to have knocked over this building in a rural area of the county.

(Courtesy of Mountain Statesman)

Paw Paw to hold aid concert

Nov. 21
The Morgan Messenger
Berkeley Springs

Heard of "Live-Aid" and "Farm-Aid"?

What about "Band-Aid"?

"Band-Aid" is a benefit variety show for the flood victims of Paw Paw. It will be held on the 14th of December at 7:00 p.m.

Local entertainment is being rounded up, and a good show is expected.

Admission will be $2.00 for adults and $1.00 for students. Children and acts will be free.

All proceeds will go directly to people suffering loss.

ALDERSON—Even dog houses weren't sacred to the floodwater from the Greenbrier River in Alderson. (Courtesy of Huntington Herald Dispatch)

LEWISBURG—An Army National Guardsman catches a few minutes of rest at the Greenbrier Lewisburg airport after working round the clock aiding flood victims with supplies. (Courtesy of Huntington Herald Dispatch)

ALDERSON—The Greenbrier River spilled into downtown Alderson. Many trailers and small buildings were swept off their foundations. (Courtesy of Huntington Herald Dispatch)

CLARKSBURG—Carrie Skinner, 86, of 412 Hart St. sorts clothes Nov. 6 at The Mustard Seed at 140 E. Pike St. Sponsored by Clarksburg churches, The Mustard Seed has been providing food and clothing for needy Harrison County residents for about 15 years. The group's membership swelled after the flood. Volunteers have spent long days collecting and distributing food, bedding, clothes and cleaning supplies to flood victims. Office workers say few have worked longer or harder than Mrs. Skinner. (Courtesy of Dominion Post)

Don't come to look

Nov. 13
The Parsons Advocate
Parsons

Don't come to Tucker County to gawk!!!

We're tired. We've lost our businesses. We've lost our homes. People are being arrested for looting while we try desperately to make some kind of sense out of all this.

If you have a strong back, a shovel or a broom; if you're good with electricity; if you can cook or wash dishes or sweep floors...we'd be glad to have you.

Imagine, if you can, shoveling mud from your brand new kitchen for five hours while sightseers look through your windows, shake their heads and "Tut, tut, tut."

CLARKSBURG—Motorboats brought help and hope to those in need along South Chestnut St. (Courtesy of Clarksburg Telegram)

CLARKSBURG—National Guard troops at Clarksburg take a quick nap at the Nathan Goff Armory Nov. 8 after four days of near sleepless duty because of flood conditions. (Associated Press Wire Photo Courtesy of Huntington Herald Dispatch)

SALEM—The Halls Run Bridge buckled from the force of high waters in Harrison County. (Courtesy of Herald Record)

SHINNSTON—South Pike Street was a sea of misery when the West Fork River overflowed. (Courtesy of Shinnston News)

"We floated down stream about three miles on a section of roof, then it overturned and I lost hold of her."

—— Elkins resident and flood survivor Bill Painter describing the last time he saw his wife

Flood damages approach $1 million

Nov. 11
Braxton Citizens' News
Sutton

The raging floodwaters which ravaged many areas of the state early last week have resulted in preliminary damage estimates in Braxton County that exceed $750,000, according to a spokesman for the local Emergency Services Council.

No serious injuries were reported as a result of the flooding, but several area families have been left homeless in its wake. "While an accurate count of households affected is presently unavailable, several Braxton Countians have lost their homes and personal property," noted Wayne Given, Director of the Braxton County Emergency Services Council.

Damage assessment teams from the Federal Emergency Management Agency visited the county on Friday, touring the affected areas, particularly those surrounding Burnsville and Orlando where flood damages were most concentrated, and Sunday's late night television newscasts listed Braxton among 29 West Virginia counties to be designated as disaster areas by the federal government.

The American Red Cross supplied initial aid to the victims in the form of cots, bedding, and food, and temporary sleeping quarters were made available at the Braxton County National Guard Armory, Braxton County Memorial Hospital, and the Burnsville Lions Den. Local volunteers throughout the county worked many long hours, often in very uncomfortable conditions, to aid their neighbors with food, clothing, and temporary housing, and to assist in the on going salvage and clean-up efforts as well.

Given also noted that he had spoken to a representative of the West Virginia National Guard on Friday to request assistance in the clean-up of debris that remained in the wake of the flooding, and that the request had received an affirmative response. "As soon as a unit becomes available they will be on hand to assist in the clean-up," Given noted.

"I'm truly saddened by the overwhelming losses suffered by the flood victims," Given concluded. "And I hope that the lives touched by this disaster can be quickly put back together. I do feel, however, that we here in Braxton County were fortunate—and should be thankful—in that no lives were lost in the flood."

Burnsville and Orlando proved to be the hardest hit areas of Braxton County, as the torrential rains raised the level of streams in that area to approximately 12 feet over normal flow. The Burnsville Dam reservoir, which serves a 165 square mile drainage area, crested at 808.80 feet above sea level at 9:30 p.m. Wednesday, while a 7:30 a.m. Monday morning reading indicated a level of 781.72 feet above sea level. The previous maximum level at the Burnsville Dam, recorded in June, 1981, was 805.01 feet above sea level.

According to a spokesman from the Corps of Engineers, the comfort station and beach area at Bulltown were flooded, and the marina comfort station was covered. The check-in station and picnic shelters at Riffle Run were flooded, as well.

The Sutton Dam reservoir, which crested at 959.1 feet above sea level, rose from a 7:30 a.m. Monday morning reading of 906.85 feet above sea level. Summer pool at that facility averages between 922-925 feet above sea level, while winter pool is 895 feet above sea level. The previous maximum level at that reservoir, in June, 1974, was 950.21 feet above sea level.

The Bakers Run campground and the Gerald R. Freeman campground were closed due to high water, as were the Bee Run and South Abutment areas of the dam.

According to John Cox, Soil Conservation Service spokesman, the five Salt Lick Watershed Dams functioned just as they were designed to. Cox noted on Thursday that the dams were all within five or six feet of being full, and were discharging water at the rate intended. "While the levels were some 20 to 25 feet above normal, none of the emergency spillways at the dams were required to func-

BURNSVILLE—When the recent flooding claimed access bridge to Mary Myrtle Edward's home, friends Chester Huffman, Rick Huffman, and Denzil Wine immediately set about the business of replacing it. "She's in her eighties, and doesn't come and go much," Rick noted, "but she likes to get her mail regularly!" (Courtesy of Braxton Citizens' News)

tion," Cox stated. "The reservoirs have all crested, and have started draining back down to permanent pool."

The Burnsville School received some water damage, and was closed on Tuesday due to the high water. Approximately 600 additional students throughout the county could not attend school due to washed out bridges, submerged roads, and other flood-related problems. By Wednesday, school attendance had almost returned to normal.

Approximately 35 mobile homes located at the Waldron Trailer Park, on Route 5 near Burnsville, received water damage as Salt Lick Creek rose to a level almost two feet deep inside some of the homes.

Two apartments located beneath the Burnsville Qwik-Pik were submerged in almost six feet of water.

The Burnsville Go-Mart received almost two feet of water, while other Go-Mart stores owned by Gassaway businessman John Heater—in Weston and Glenville suffered much higher losses as they were inundated by the fast rising waters.

Many residents of the area were left without gas and electric service, and others lost access bridges to the rapid-flowing, swollen streams.

According to Dr. Ralph Stalnaker, ER physician at Braxton County Memorial Hospital, the Department of Public Health had ruled on Thursday that routine typhoid immunizations would not be necessary, but advised persons receiving cuts or injuries while working in flood-damaged areas to check with their family physician or a local health care facility to ascertain whether a tetnus shot should be administered.

BRAXTON COUNTY—When floodwaters receded from the Burnsville and Gem areas, a panorama of damaged riverside properties such as this were discovered. (Courtesy of Braxton Citizens' News)

BUCKHANNON—Rowboats had to be used to rescue many flood-trapped victims from their island homes in Buckhannon. (Courtesy of Record Delta)

BUCKHANNON—Not everyone is upset with the rescue process. (Courtesy of Record Delta)

CAIRO—Floodwaters are beginning to recede in downtown Cairo. (Courtesy of The Parkersburg Sentinel)

Flood aid called 'real cruel hoax'

Nov. 15
Dominion Post
Morgantown
By P. Kay Nottingham

Del. Larry Schifano, D-Monongalia, Thursday night called it a "real cruel hoax" to tell flood victims help is available from federal and state agencies.

"The help isn't there if those people build on the flood plain," Schifano said.

Del. Liz Martin, D-Monongalia, said she will address the Legislature Sunday on what she considers a "vicious trap" where people are told centers are available when filling out applications may not mean people will get assistance.

She said a small percentage of people will receive federal assistance.

"That bothers me," Martin said.

Schifano and Martin were among the handful of people attending a talk session Sen. Stephen Cook, D-Monongalia, set up at the Monongalia County Courthouse Thursday to hear problems and questions from people affected by flooding.

CAIRO—The North Fork of the Hughes River overflows blocking the road into Cairo. (Courtesy of The Parkersburg Sentinel)

Elk & Gauley crest at highest marks in 50 years

Nov. 6
Webster Echo
Webster Springs

Property damage in Webster County has been estimated to run into the millions of dollars, as flooding occurred on Monday, November 4, and the waters of the Elk and Gauley Rivers crested to their highest marks in over 50 years. Numerous people left their homes voluntarily to seek higher ground, or were evacuated by emergency personnel to places of safety during the evening and night; however, to date, no fatalities nor injuries have been reported in Webster County, according to officials.

Serious flooding was also reported in 33 West Virginia counties, as well as in parts of Virginia and North Carolina, in the wake of the remnants of Hurricane Juan.

Schools in Webster County closed early Monday afternoon, as the water began to rise, caused by heavy, steady rain. By late afternoon, the Elk was already surging over its banks in some places, and the Gauley had rendered some roads in that area impassable. One small busload of school children was forced back from the Dyer when the driver tried to take students home, the Williams River rendering the road impassable; children returned to their school and notified parents and made other arrangements.

Before nightfall, Elk River was spilling rapidly into Baker's Island and the areas of Webster Springs, Bergoo, Curtin, Parcoal, Cherry Falls were also in a full flood.

At 6:17 p.m. the electrical transformer, located at the bus garage site in Webster Springs, was well underwater, with the river reaching up to the transformer's controls. At this point the Monongahela Power Company made the decision to cut power off then, as a safety measure. A spokesman explained that the risk of fire was present. On Tuesday, the spokesman explained that, although the power was turned on again at 5:50 a.m., extensive damage had been done to the station, including the fence around it having been destroyed. The station was kept under surveillance all night, and by Tuesday morning all customers had been returned to power except for approximately 10-15 in the Curtin (Barton) area.

The Webster Springs Volunteer Fire Dept. was joined by their counterparts in the Diana V.F.D., and along with the Sheriff's Dept., State Police, and city police, worked throughout Monday night and Tuesday along the Elk. A great many persons were evacuated from their homes in the Parcoal, Cherry Falls, Curtin, Dorrtown and Bergoo areas, some taken to the Webster County Senior Citizens building and some to Webster County Memorial Hospital for safety. The rescue crews also took boats to some homes in the Parcoal area, recovering valuables. Also out in force to battle the raging Gauley were members of the Cowen and Erbacon V.F.D.'s.

The Gauley River crested at 1:30 a.m. Tuesday, reaching 23 1/2 feet, one-half foot over the flood stage. Rainfall in Camden-on-Gauley in the 24 hour period yielded 4 1/2 inches of rain. In Webster Springs, rainfall for the same 24 hour period was 4.74 inches.

Areas of the county most affected were Webster Springs, Parcoal, Dorrtown, Curtin, Bergoo, Williams River, Gauley, Camden, Bolair, parts of Black Fork and Cherry Falls. Also, minor flooding was reported in the Birch River area.

A spokesman for Elk River Sewell Coal Co. in Bergoo stated that their railroad transportation line for the coal has been severely damaged, with repairs expected in about 3-6 weeks. One small railroad trestle was completely washed away in the flood, plus 200-300 feet of railroad track were also washed out. The Board of Education office in Webster Springs sustained only minimal damages, said a spokesman, with a window in the basement broken out and water pouring in to about 4 feet high.

In Bergoo, Ron O'Brien's trailer park was badly damaged, suffering between $40,000-$50,000 worth of damages, as six trailers were seriously damaged by the flood.

BERGOO—The Western Maryland Railway line twists into the murky Elk River after being undercut by the swift-moving water. (Courtesy of Webster Echo)

WEBSTER SPRINGS—Mayor Cassandra Given inspects damage to Baker-Addison Pool. (Courtesy of Webster Echo)

PARCOAL—The Clutter home was one that received extensive damage. (Courtesy of Webster Echo)

CHERRY FALLS—The shelter was washed from its foundation and remains some feet below. Note debris alongside. (Courtesy of Webster Echo)

WEST UNION—The water treatment plant becomes an island in Middle Island Creek. (Courtesy of Herald Record)

WEST UNION—The Middle Island Creek overflows in West Union. (Courtesy of Herald Record)

RIVESVILLE—Children attempt to forge down a familiar street in unfamiliar circumstances. (Courtesy of Fairmont Times)

WORTHINGTON—An unidentified man examines damage to this Marion County town. (Courtesy of Fairmont Times)

Another type of war zone

Nov. 20
The Morgan Messenger
Berkeley Springs

The following is something I felt like saying for Veterans Day.

It's just getting light enough to see something besides movement. About 30 meters back there's some loud voices and a deuce and a half pulls around a bunch of friendlies with a couple of throttle pops. The stink of diesel fuel and muck hangs in the air. A lot of guys are wandering around about half fogged out in grubby clothes, getting the job done, but slow. I'm ready to go down for a week myself, but I know that ain't gonna happen any time soon.

Over our head there's a whop-whop-whop and a Huey with big red crosses comes in from the south, while a fighter cuts across from the East low enough to be gone in a hurry. I wonder who the hell is gonna show up with the problem for today or are they just gonna let me pick through my crap in peace.

For the time just described my body is in the aftermath of the Moorefield flood, but my mind is back in Qui Nhon, An Khe, Plieku, Charang and Phutai. Today I'm glad I went to those places because this part is easier the second time around.

——guest editorial in *The Moorefield Examiner*

WORTHINGTON—The floodwaters had begun to recede along State Rt. 19 as Nov. 5 wore on. (Courtesy of Fairmont Times)

WORTHINGTON—Bridges such as this were of no use to residents when the floodwaters reached record heights. (Courtesy of Fairmont Times)

TYLER COUNTY—Clothes and rugs hung on a line escaped the floodwaters of Middle Island Creek though 30-40 homes in Middlebourne were flooded. (Courtesy Tyler Star News)

TYLER COUNTY—Wildlife throughout the Mountain State suffered from high waters and flash flooding, as the body of this deer washed up along W.Va. 18 clearly indicates. (Courtesy Tyler Star News)

MIDDLEBOURNE—Floodwaters from Middle Island Creek swelled across W.Va. 18 in the Tyler County seat as 80-100 families suffered from the flooding. (Courtesy of Tyler Star News)

BIG BEND—Trying to save clothing and other valuable items as the water came up, Tim Blair, left, and Rex Yoke carry a box from a mobile home park which was later flooded. (Courtesy Parkersburg Sentinel)

High water concerns area

Nov. 6
Wirt County Journal
Elizabeth

Wirt Countians spent Nov. 5 preparing for the full impact of several days of rain.

Reports of flooding "up the river" (the Little Kanawha) and the anticipated crest of the river in various parts of Wirt County were fed into the individual decision making process of citizens, many who had to decide whether or not to move possessions from homes and or businesses.

The LKR was expected to crest at 38 feet in Elizabeth late Nov. 5 or early Nov. 6. Flood level in the county seat town is 36 feet.

The most frequently asked questions were, "Are you going to move?"—"Will the river get as high as in 1967?"—"Can I help you?"

A communication center was established at 2 p.m. on Nov. 5, at the Elizabeth-Wirt Fire Department and representatives of the state, county and town planned to be on duty throughout the night.

Health officials were on call and they were coordinated by the Wirt County Health Department.

Authorities were prepared to assist citizens in all types of emergencies.

Wirt County schools were closed early on Nov. 5 and all day Nov. 6.

Several highways and roads were reported under water at 7 p.m. on Nov. 5, including Rt. 14 near Palestine, Rt. 6 near Newark and Rt. 47 near the IOOF camp.

Mayor calls emergency city session

Nov. 6
Clarksburg Telegram
Clarksburg

A majority of Harrison County residents remained without water service today, as officials at midnight Tuesday ordered a complete shutdown of water to all customers of the Clarksburg Water Board.

"The situation's not good," said Water Board Manager H.E. Cain, who noted that water tank supplies had dwindled to just a matter of hours. Officials Tuesday night ordered the complete shutdown in the event water is needed for a major fire.

Meanwhile, an emergency session of Clarksburg city council was called for 9 a.m. today by Mayor Albert N. Cox II regarding the flood problem situation.

"The city is in a state of emergency," said Cox. "Gov. Moore has issued an executive order for all banks, directly serviced by the Clarksburg Water Board, not to open today."

The mayor stated that, at latest report, there was still four feet of water in the water filtration plant building on South Chestnut Street.

GRANTSVILLE—Not on a pleasure ride, this motorcyclist fords the flood on Rt. 5 going toward Glenville. (Courtesy of The Calhoun Chronicle)

GRANTSVILLE—Water was still rising at the rate of four inches per hour as this Army Reserve truck plowed through the floodwaters of the Little Kanawha River on Nov. 6. (Courtesy of The Calhoun Chronicle)

BIG BEND—As the Little Kanawha rose this homeowner slogged through the muck to salvage rugs and furniture from his Calhoun County house. (Courtesy Parkersburg Sentinel)

WIRT COUNTY—This Creston residence appears ready for visitors to "sit a spell." (Courtesy of Parkersburg News)

CRESTON—Wirt County volunteers prepared to launch a recovery effort during the flood (Courtesy of Parkersburg News)

56 soil conservation projects let to reclaim damage, bury livestock lost by S. Branch valley farmers

Nov. 14
Clarksburg Exponent
Clarksburg

The Soil Conservation Service of the U.S. Department of Agriculture has contractors working on 56 jobs to help West Virginians recover from devastation caused by flooding and avoid further damages.

SCS technicians are supervising flood reclamation costing nearly $3 million in seven of the hardest hit counties, according to State Conservationist Rollin N. Swank. Another $1 million was obtained on Tuesday for additional work.

Swank also reported these other developments:

He is seeking additional millions to help alleviate danger and suffering in West Virginia due to destroyed stream channels and other sediment-related problems from the floods. The national office of SCS provides 100-percent federal funding, enabling the SCS organization in the state to relieve imminent hazards to life and property.

Employees in the state office in Morgantown worked through the weekend, issuing and administering the 56 contracts which were let by the end of the day Tuesday. Contractors were notified by phone, radio and personal contact due to the immediate need for the work.

In the South Branch Valley, where thousands of cattle and millions of chickens and turkeys were killed by drowning, the SCS is coordinating burial of dead livestock, working with the National Guard. The SCS has identified burial sites. At least two soil conservation districts from elsewhere in the state, the Tygarts Valley and the Western, have sent in heavy equipment to bury livestock.

SCS technicians have evaluated all of the 64 flood-prevention dams the agency has built in the last 30 years in the Potomac branches and forks flowing north through West Virginia's eastern counties. Swank said every dam functioned as designed, to retard flood waters, and no SCS dam failed.

He said the flood-retarding dams helped save some communities from severe flood damages, such as Maysville and Keyser. In other places, such as Petersburg and Moorefield, the flooding was beyond the capacity of any manmade structures to control, he said.

"Our dams are built to hold back the runoff from a storm that would occur only once in 100 years," he said. "We have an emergency spillway beside each dam, so the water will not over-top the dam, but water would not flow through that spillway unless a storm exceeded the 100-year event.

"On 21 of our 64 dams in the headwaters of the Potomac, flood water flowed through the emergency spillways. That was true on nearly all the dams in Pendleton County. On the big dam on Patterson Creek in Grant County that was celebrated as the 100th dam in the state in 1970, water was still flowing through the emergency spillway on the Wednesday following the generalized flooding Monday."

The SCS-contracted work is to correct "watershed impairments" brought about by the flooding, which pose an immediate threat to life or property. Swank said streams are flowing through and next to homes and businesses and must be relocated so residents can safely rebuild their lives.

Examples are homes ready to fall in creeks or complete blockage of streams or bridges.

"The amount of sediment deposited and land scoured out by the flood is almost unbelievable," Swank said. The emergency funds being used by the SCS may not be used to repair private or public transportation facilities, public utilities, or similar facilities."

Aerial view of flood sites reveals nature's ravages

Nov. 13
Clarksburg Telegram
Clarksburg
By Myra Litherland
Telegram Staff Writer

The floodwaters which devastated a 29-county area of West Virginia last week have receded.

Left in their wake, however, was a death toll of 28 and another 28 missing, more than $220 million in damages and some 8,000 families homeless, by Monday's accounting.

An aerial view of many hard-hit areas of the Mountain State reveals scenes not unlike a bed-ridden child's Monopoly game gone awry.

The soft mauve-gray hills and hollows, flecked by the brilliant golds and oranges of late-autumn foliage, the wide expanse of green farmland and the blue-gray strip-mined swaths, resemble a rumpled quilt.

Brindled ribbons of water negotiate a multitude of hairpin turns throughout the mountains.

Various communities, large and small, hug the banks of those rivers, streams and creeks; white-frame homes, red-brick churches, business-lined streets—and few of them escaped the wrath of last week's deluge.

Some tributaries, rain-swollen and able to remain within their banks no longer, ignored their winding boundary and just raced across, taking with them everything in their paths. Others widened, inundating anything nearby before spreading out to swamp surrounding areas, their receding waters leaving only brown-marbelized expanses of desolation behind.

Some structures are completely gone—only the foundations are left. Others are flattened or severely damaged. Still more are shoved into odd angles, as if the child's "game board" had been tilted.

Entire mobile home courts have disappeared, while bridges which once spanned the waters exist no more, as do the roads leading up to them.

Scattered over all is the matchstick-stacked array of debris which once constituted trees, buildings, automobiles, homes, family possessions.

Those West Virginians more fortunate than their flood-stricken neighbors have opened their hearts and pursestrings in an outpouring of volunteer help and aid to the afflicted areas. Money, meals, cleaning supplies and clothing have poured in, almost equalling the velocity of last week's careening waters.

Although a lack of clean water remained a problem in some areas Monday as did clogged sewer lines, downed power lines and erratic telephone connections, cleanup efforts are underway.

The process will be lengthy, stretching over many months, and some areas may never be the same again.

But other flood-wracked parts of the state will be rebuilt and renovated, and life will return to normalcy.

A flight over the hills of West Virginia will once again offer a breathtaking view.

"Everybody has a flood story."

—— Headline from the Buckhannon Record Delta

Epilogue: 'One clear call for me'

Coral Gables

Mark Zimmer, a meteorologist with 27 years' experience working for the National Hurricane Center or related agencies, said Hurricane Juan's influence was the "major contributing factor" to the record rainfall received in West Virginia during the days preceding the Killing Waters.

Juan was tracked as having passed through the Ohio Valley and on up in the area of the Great Lakes before he died for good.

But the moisture Juan had brought to the area collided with another disturbance, Zimmer said, and resulted in greater amounts of rain than either system alone would have dropped. In that respect, Juan ended jointly with another storm, just as he had begun with Xina.

"Juan," Zimmer said, "was a half-breed."

To the end.

★★★★★

Albright

Maxine Wotring was asked what she missed the most.

She cast her sad eyes toward Jack. "Him a playin' his guitar," she said.

When the Cheat River invaded their home in the early hours of November 5, they had fled first to the porch. "Go on!" Jack had shouted to her, "I have to get some things."

But Maxine was resolute in her demand for him to come along with her. "I don't want to lose you, too," she said.

Move on, move on!"

They escaped.

But their home had completely vanished in the flood waters.

Everything, including Jack's favorite guitar, other instruments, a gun collection and several of their heirlooms—lost down the river.

"We were standing up there on the hill that morning watching our house go down," Jack commented. "I just threw up my hands and said, 'Well, hell, there it goes.'

"But I'm quite sure that's the story of all old people who lived there. Things that you'd intended on passing down to your grandchildren and their grandchildren went down the river.

"But I never question God's actions at all. If this is His wish or desire or whatever you want to call it, why, He's always blessed me quite well. You have to give thanks for what he has taken away . . . and for what He's left you. He always manages to just leave you what you need."

The Wotrings plan to re-build in Albright.

"Life's got to go on," Maxine said.

"That's just it," Jack added. "We have some idea just how uncertain we are on this here little ole' crust that we live on. But as long as you're living, you simply pick up what you have and go on about your business. I'm all the time planning for that tomorrow, you know. Yesterday, it's done gone by. Hell, you can't worry about it. Just pick up and go ahead and make out of life what you can.

"A lot of people said to me, 'I bet you've never seen anything like that,' and I said, 'No, I've seen worse. I landed on Omaha Beach and they (the officers) kept yelling, 'Move on! Move on!' Now, that was rough."

A little while later, he added, "You can't get too smug. You got to get woke up once in a while. Now, I've been woke up several times," he laughed. "This is enough."

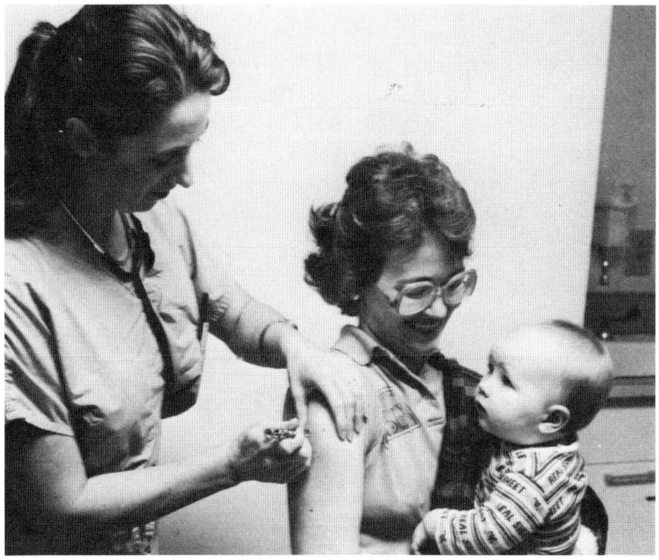

GRANTSVILLE—Gina Gherke, center, gets a diphtheria/tetanus shot from Nurse Dawn Fox while she holds her 10-month-old daughter Jessica. (Courtesy of The Calhoun Chronicle)

★★★★★

Country Club

A lot of the bells are gone.

A favored doll collection, too, washed away, but one of the most delicate of the dolls was found unblemished a few miles downstream in Albright, Dolly said.

It is one of the few material things she now has to remember Lynn.

No one knows for sure what happened that night at Lynn's cottage.

Dolly's sons did manage to find one of several antique clocks which Lynn had collected. The hands had stopped at 4:17.

Dolly seemed convinced that Lynn's streak of stubbornness wouldn't let him abandon his place. "Lynn, I think," she said, "made his own choice.

"He had cancelled his flood insurance two years ago," she added, "because, as he told me, 'It (the Cheat) won't come up, I know it.'"

During a phone interview, Dolly reflected on her love for her brother. "Oh, he was feisty, all right, but it was just his cantankerous little quirk and everybody loved him for it. In spite of his toughness, he was very, very soft. Some of the most precious times we had were during our farewells and our greetings. He always broke down, and it was so touching to me because other times he seemed to be so tough.

"He had written down on a piece of yellow paper the things he wanted done after he died. 'No funeral, but if so, quiet,' he wrote. 'No music, no singing, no viewing, no flowers, no obituary, and just one spray of flowers on the coffin.'

"At the end of that little piece of yellow paper he had written one of his favorite poems. Oh, he was always reading poetry, always reciting it. But this one proved to be so ironic."

It was from Tennyson's **Crossing the Bar**

"Sunset and evening star,
And one clear call for me!
And may there be no moaning of the bar,
When I put out to sea."

"Parsons is down, but not out!"

—— Rev. Meade L. Gutshall

Additional copies of the second edition of this book may be purchased by sending $12 plus $2.50 for shipping and handling for each book to Cheat River Publications, Inc., 805 East State Avenue, Terra Alta, WV 26764.